MANUFACTURING TECHNOLOGY

Advanced Machine Processes

MANUFACTURING TECHNOLOGY

Advanced Machine Processes

H C TOWN and H MOORE

BATSFORD ACADEMIC AND EDUCATIONAL LIMITED
LONDON

ISBN 0 7134 1096 5 (cased edition)
     0 7134 1097 3 (limp edition)

Typeset by Tek-Art Ltd London SE20
Printed in Great Britain by
Billing & Son Ltd
Guildford & Worcester
for the publishers
Batsford Academic and Educational Limited
4 Fitzhardinge Street, London W1H 0AH

CONTENTS

1

GEAR CUTTING MACHINES AND PROCESSES

HISTORICAL DEVELOPMENTS

The mathematical theory of gear design has a lengthy history. Robert Hooke in 1666 stated in his *Principles of Gearing* that 'gear wheels should have as many teeth as possible', that 'forces and rotary motions should be constant' and that 'the point of contact between the teeth should always lie on the line that joins the centres'. The theoretical aspect of tooth form was being studied as early as 1451 by Nicholas of Casa who recommended the cycloidal curve, but the French mathematician, Phillippe de Lahire, in 1694 came to the conclusion that of all tooth forms the involute was the best. Unfortunately the cycloidal shape received more support, and the involute curve was not adopted until well into the nineteenth century.

As with the milling machine, development of the gear cutting machine was retarded by the lack of suitable cutters, but J G Bodmer built a machine in 1820 using a fly cutter for producing wooden gears and was the first engineer to use a true formed cutter on metal gears. Another engineer, Richards Roberts of Manchester, in 1821 advertised in the *Manchester Guardian* that he 'Respectfully informs iron founders, machine makers, and mechanics, that he has Cutting Engines at work on his new principle, capable of producing any number of teeth required, the teeth will not require filing up'. In 1844 Joseph Whitworth joined the pioneers by building a large machine which produced gear teeth by a formed milling cutter. Accurate tooth spacing was obtained from an index plate with rows of holes to suit the number of teeth.

THE FORMED CUTTER SYSTEM

The first major improvement towards accurate gear cutting came in 1864 from the Brown & Sharpe Co, USA. The difficulty with existing cutters was that sharpening was virtually impossible. The improved cutter (Figure 1.1) had segmental teeth, each relieved on the outside diameter with the flanks conforming to the contour required. Thus sharpening was effected simply by grinding the front face of each tooth, which did not change the contour.

The shape of a gear tooth changes with the number of teeth in the gear,

Figure 1.1
A form relieved gear cutter

varying from pronounced curves on a pinion with, say, 10 teeth, to the straight sides of a rack. This means that a set of cutters must be provided to cover all numbers of teeth for a given pitch. Because of the more complicated shape of a cutter with cycloidal curves, 24 cutters are required for every pitch, but only eight for the involute shape.

ELEMENTS OF TOOTH ACTION

The conditions for the correct running of gear teeth are that the ratios of the speeds or angular velocities of the mating gears must be the same at every point; thus the common normal at all points of contact must pass through the pitch point P (Figure 1.2). The elements are based upon the

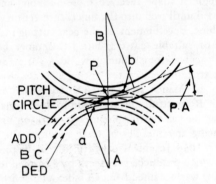

Figure 1.2
The elements of gear tooth action

pitch circle which is used as a basis for calculations. With the involute gear a single curve only is employed; this reaches from the addendum to the base circle and thence continues by a straight radial flank to the bottom of the tooth, where it terminates by a small radius for strength. If A represents the driver and B the driven gear, then aPb is the path of contact for the direction shown. If a tangent is drawn to the pitch circles, the path of contact makes this tangent into an angle which is termed the pressure angle of the gears (PA). The shape of the tooth depends upon the number of teeth in the gear and the pressure angle, which is usually $20°$ or sometimes $14\frac{1}{2}°$. The path of contact is terminated at the points a and b by the addendum circles.

The tooth action is determined when the driving tooth A commences engagement with B at point a. At this instant the relative velocity of the teeth at the point of contact is largely sliding and friction loss a maximum. The period a to P is termed the 'approach' contact and the velocity of sliding decreases uniformly, becoming zero at the pitch point P where a rolling action takes place. The period P to b is termed the 'recess' contact, and the velocity of sliding reverses direction and increases from zero to a maximum at b where contact ceases.

The importance of a large pressure angle is shown in that undercutting commences when the number of teeth is less than 28 for a pressure angle of 14½° and less than 15 for a pressure angle of 20°. Thus we need more teeth for greater strength, until the rack tooth is reached, when the involute becomes a straight line. Undercutting can be corrected in three main ways: (1) by using stub teeth and thereby reducing the working height (this gives a strong tooth but durability is lessened) (2) by altering the pressure angle so that the interference point is moved; (3) by correcting the addenda. This is a method with generating systems in which it is possible to alter the addenda and dedenda of both gear and pinion while keeping the height of tooth standard. This is effected by enlarging the outside diameter of the pinion and reducing that of the gear by the same amount: Figure 1.3 shows how any part of the involute curve may be produced at any pressure angle. A is the usual shape for a 12-tooth pinion,

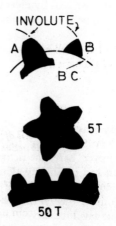

Figure 1.3
A diagram showing the effect of pressure angle on involute gears

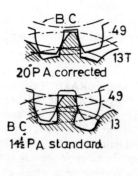

Figure 1.4
The result of undercutting and means of correction
BC is base circle

of pressure angle 14½°, whereas B has an increased addendum and increased pressure angle to use more of the involute than A does. Both teeth were cut by the same Sunderland cutter, and the same applies to the

5-tooth pinion and 50-tooth wheel adjacent. Figure 1.4 shows the effect of the correction methods (1) and (2) on the meshing of another pair of gears.

GEAR GENERATING MACHINES

The theory of generating gear teeth was established long before practical applications were possible. Edward Sang, Professor of Mechanical Philosophy in Constantinople, suggested in 1837 that the involute form should be employed for gear teeth and he enunciated the law for tooth contact; but he went further for, in a remarkable address before the Royal Scottish Society of Arts, he suggested how gear teeth could be generated long before the practical means were available.

Sang stated, 'Having made a cutter with a rack as outline, and mounted it on a carriage with a traverse screw, the cutter being arranged so that its pitch line may pass the proper distance from the axis of the index wheel, make a series of cuts all round. Then we shall have a wheel whose teeth are truly formed to work with our assumed rack, and all wheels formed in this way will gear truly with each other.' Still Sang went on, '. . .instead of having the rack as our prototype, we might have assumed any one wheel. Thus, having obtained a pinion, we may be required to cut the teeth of a wheel with it. Let the assumed form of the cutter be notched to form a series of cutting edges, and let it be fixed in such a manner as to slide parallel to the index wheel while it be turned upon its own axis by means of the index plate. Then is the blank toothed in such a way as to gear truly with the pinion.'

Sang obviously saw the Sunderland rack cutter and the Fellows pinion cutter in an age when generated cut gears had not been thought of.

TOOTH GENERATION BY RACK-SHAPED CUTTER

In the year 1908 (ie 71 years after Sang's theory was propounded) a Keighley engineer, trading under his own name of Sam Sunderland, took out a patent which made his name known the world over. Later the rights of production were acquired by J Parkinson & Son Ltd of Shipley. The success of the rack process is based upon the principle that, since every wheel will mesh with a rack of the same pitch, it follows that, if the cutting tool is made in rack form, it will generate a true involute curve. (See Chapter 11, Figure 11.43.)

In operation the blank is rotated uniformly about its centre and simultaneously the cutter is caused to advance along its length, the speed of advance being equal to the speed of the pitch line of the blank. After the cutter has advanced a distance equal to one or more pitches, the motion of the blank is arrested, while the cutter returns to its starting point. The cycle is repeated for every tooth so that all are generated under the same conditions.

Figure 1.5
A Sunderland gear generating machine
(by courtesy of J Parkinson & Son Ltd, Shipley)

SUNDERLAND GEAR GENERATING MACHINES

Figure 1.5 shows a machine for cutting gears up to 1520 mm in diameter, a face width of 300 mm and module 18, but larger machines have been built for years which are 4600 mm in diameter with a face width of 710 mm. The gear is mounted on the left-hand unit with its axis horizontal, this lay-out allowing both large wheels and small pinions to be cut on the same machine. The cutter traverses in its guides across the face of the blank at a speed and feed selected by the levers on the side of the body, while rapid traverses and in-feed are by hydraulic operation. Electrical controls include a push button panel and selector switches on a pendant control, while the control elements on the base plate provide an automatic cycle which, once set up, enables a complete cycle of operation to be carried out automatically, requiring only loading and unloading of the component. The cutter boxes hold a patented self-compensating cutter which can be removed for sharpening and which can then be replaced without having to reset the machine. Double cutting can be carried out, two cutters being mounted back to back, one cutting the sides of the tooth on the forward stroke and the other on the root of the tooth on the return.

RACK CUTTERS

A pair of double-acting cutters with a support plate is shown in Figure 1.6(a)

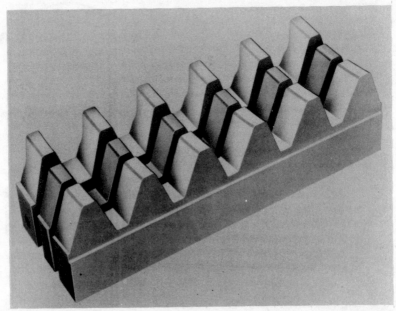

(a)

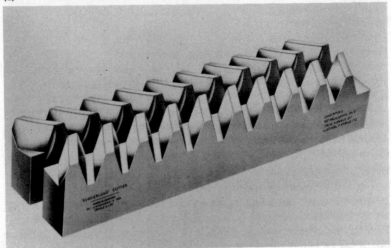

(b)

Figure 1.6
Types of rack cutters used on the Sunderland machine

whereas Figure 1.6(b) shows a pair of cutters for machining double helical gears. No special equipment is required for cutting single helical gears, for the cutter slide is simply set to the correct helix angle. For cutting double helical gears a special saddle with two slides inclined at a fixed angle is used. The right- and left-hand sides of a tooth are cut simultaneously; as one cutter approaches the centre on the cutting stroke, the other is withdrawn. The cutting edges come to rest when they just pass the central plane dividing the two helices and in this way the space formed by one cutter is clearance for the other. The facility to produce double helical gears without a gap enables gears of maximum strength and load-carrying capacity to be designed. No radial contour is produced as with double helical gears produced with end milling, a process sometimes used for large gears.

There is no alteration in tooth form when rack-shaped cutters are reground, and, by using support plates, cutters can be sharpened until they are only 5 mm thick. The straight sides of the cutter facilitate ease in manufacture and reduce the cost so that some hobs are five times and some pinion-type cutters twice as expensive. High rates of metal removal are possible and even the largest pitches can be generated. Cluster gears can be cut, and by employing special cutters the process can be extended to the cutting of ratchet wheels, chain sprockets, and gears with stub teeth.

TOOTH GENERATION BY PINION-TYPE CUTTER

The pinion-type gear generating system was the first to be established 60 years after Sang's theory. In 1897 E R Fellows founded the Fellows Gear Shaping Co in Springfield, USA, and built the first machine in 1900. The system is based upon the feature that, as a pinion will gear with any wheel of the same pitch, a pinion-shaped cutter will produce involute teeth if cutter and blank are rotated together after the cutter has been fed in to depth. The success is based upon the peculiar property of the involute, ie the possibility of varying the centre distance while retaining uniform velocity transmission.

The machines using the system are built in a variety of types, both vertical and horizontal. The machines are very versatile and are used to cut not only straight spurs but also single and double helical gears with a central space; in addition they are the only machines which are used to cut internal gears in any quantity. Conjugate shapes, as shown in Chapter 11, can be produced. Only one cutter for all gears of the same pitch is required, and the versatility of the system can be seen from Figure 1.7. This shows a single-tooth pinion meshing with a 63-tooth wheel. The minimum number of teeth that can be used for a spur pinion is three, but by using the double helical form the one-tooth pinion was cut on a standard machine using standard cutters for both wheel and pinion. The proportions are as in Table 1.1.

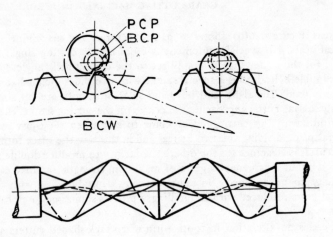

Figure 1.7
A single-tooth pinion in mesh with a wheel of 63 teeth

Table 1.1

	Pinion		Wheel
Pitch	1.986 (diametral)	or	1.595 (circular)
Pitch circle diameter	0.508 in.		32.004 in.
Outside diameter	1.756 in.		32.210 in.
Face width		6 in.	
Centre distance		16.26 in.	
Pressure angle		22° 10'	
Ratio of pitch diameters		63 to 1	
Ratio of outside diameters		18.28 to 1	

As the teeth are helical and half the face width is equal to the lead of the pinion, there is continuity of contact giving uniform and positive transmission, and the abnormal outside diameter of the pinion ensures sufficient power, actually transmitting 10 hp at 100 rev/min.

Nearly all tooth contact takes place in 'recess'. This is preferable, for the action takes place as the pinion is going out of engagement and thus oil is drawn into the line of contact and provides efficient lubrication.

On some vertical machines the pitch that can be cut is limited to a diametral pitch of about 4 owing to the overhung cutter, and for cutting helical gears a limitation is that a separate helical guide is required for every change in lead as well as a special cutter. Figure 1.8 shows a diagram of the construction indicating the reciprocating motion as the cutter and blank rotate in unison while the helix on the guide at the top of the spindle provides the lead.

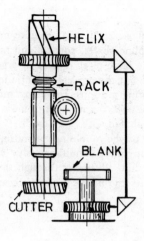

Figure 1.8
A diagram showing the action of a pinion-type gear machine

PINION-TYPE CUTTERS

Owing to the necessity of providing relief to the cutting edges, the diameter of the cutter decreases as it is resharpened and the tooth form undergoes a change. This does not, however, prevent the formation of a constant tooth profile on the generated gear. In Figure 1.9, D represents the diameter of a

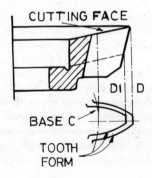

Figure 1.9
A pinion-type cutter at different stages of its life

new cutter and D_1 is the diameter at the end of its useful life. The pressure angle has changed considerably but, since the cutter and blank are rotated with the same velocity in each case, the base diameter of the generated gear will be constant if the base diameter of the cutter is also constant. The cut is taken in an advantageous way, for the edge which finishes the involute outline does the least amount of cutting.

THE HOBBING PROCESS

The hob has a very wide field of use for it will generate not only spur, spiral and worm gears but also spline shafts, squares and ratchet wheels with equal facility. In hob design a large pitch requires a large hob, which in turn reduces the rotational speed, but the process is continuous and there is no idle stroke as with reciprocating motions. Hobs cost more than rack cutters, but they do not compare unfavourably with pinion cutters, and the life is longer because the cut is shared with more teeth.

Hobbing is a milling operation of the moulding-generating type which removes metal by three motions operating simultaneously, ie the hob and the work rotate together with the hob advancing through the work, as in Figure 1.10. For cutting spur gears the hob is set at such an angle as to

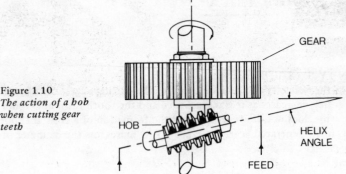

Figure 1.10
The action of a hob when cutting gear teeth

make its cutting side parallel with the axis of the gear, ie it is set at the angle of the helix, measured at the pitch line. Thus the teeth of the hob when set in this position correspond with the teeth moving endwise at constant speed into a toothed gear.

There is, however, an added complexity when hobbing helical gears. Assuming that both hob and work are stationary, it will be apparent that the hob cannot be fed in a vertical direction without the teeth fouling, so that, in addition to the uniform motion for spur gearing, it must receive a differential motion equal to the motion it would have if it had rolled across the face of a stationary gear. Hobs for cutting straight-tooth gears are usually made with a single-start thread, whereas for helical gears multi-start hobs are commonly used.

HOBBING MACHINES

Both vertical and horizontal spindle machines are available, and in general the most important member from the accuracy standpoint is the index worm gear, for any errors are directly reproduced in the work in proportion to the relative pitch diameters. Figure 1.11 shows a machine by W E Sykes Ltd operating on a spiral timing gear on a cam shaft. On this design of machine the hob only traverses backwards and forwards, so that the

Figure 1.11
A hobbing machine for cutting spiral gears (by courtesy of W E Sykes Ltd)

hob head and slide are of simple construction with low overhang and high rigidity. The working cycle is automatically controlled by a single push button, and the machine has a differential and a gear with automatic cycling and infinitely variable hob speed controls. Hydraulic work clamping and automatic work loading can be provided.

NUMBERICAL CONTROL
Numerical control (NC) to the manufacture of gears is more limited than to some other machining operations, mainly because of the geometric nature of the tooth form and the limited variety of the motions required with gear cutting during a working cycle. Figure 1.12 shows the Pfauter gear hobbing machine in which any gearing linking the hob spindle and work table is replaced by electronic arrangements which utilize encoders and dc motor drives and which can be programmed by means of digital switching. By this means the operator is able to programme the number of teeth required to be cut and the distance between the axes of the hob and the table. With this information, the control equipment will divide the

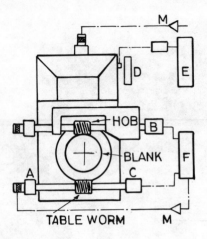

Figure 1.12
*A diagram of the
Pfauter hobbing
machine with NC*

number of pulses of the hob-head encoder by the number of the workpiece teeth to provide a signal for control of the dc motor whereby the work table is driven. A longitudinal scale D is utilized to set the centre distance between the hob and table and, with the aid of this scale, the column carrying the hobbing head is radially displaced in increments of 0.01 mm to provide hobbing depths for roughing and finishing cuts. Digital read-out equipment is incorporated to provide visual indication of this centre distance. Pfauter are also developing an NC machine incorporating more decade switching for setting the helix angle and axial travel.

The electronically controlled dc motor is shown at A coupled to the table driving worm. The measuring system consists of the encoder at the hob head B emitting 36 000 pulses/rev of the hob and a second encoder at the indexing worm shaft C, emitting 36 000 pulses/rev of the table. Two amplifiers are indicated at M while E represents the control for the centre distance and F the control for the number of teeth.

A hobbing technique introduced by Azuma of Japan comprises a negative rake carbide-tipped hob for finishing heat-treated gears where the tooth flanks have a hardness up to Rockwell 64C. Gears are first machined in the soft state to the correct root size and are then hardened. Final material is only removed after this operation and, depending on the finish required, the need for subsequent tooth grinding either can be reduced to one pass or can be eliminated.

PRODUCTION OF GEARS BY COLD ROLLING

The process offers important possibilities as compared with a conventional machining, but there are inherent limitations. Metal is displaced by heavy pressure, just as in thread rolling, so that the process is one of controlled

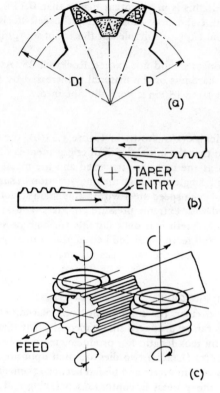

Figure 1.13
Methods of producing gear teeth by cold rolling

metal displacement by extrusion. The machines available may be divided into two types: (1) those operating with racks or with rotary dies, the axes of which are parallel with that of the blank, the process being suitable for high-pressure angle work, or (2) those operating with rotary or worm-type dies (the axes are now at about $90°$ to that of the work) the process being capable of producing much lower pressure angles and larger pitches.

Figure 1.13(a) indicates the basic action of the process in that the metal in the area A is forced to flow in the direction of the arrows to fill the area B, such that the initial blank diameter D is increased to D_1.

MATERIALS FOR ROLLING

In the process the material is stressed beyond the yield point in order to flow plastically. Work hardening can occur depending on the material analysis, the initial hardness and the degree of plastic deformation. The material should be ductile and an elongation figure of 12% or more is desirable, and it should withstand the stresses of cold working without disintegration. Many of the 'free machining' steels are not suitable since they include sulphur and, if this is in a content higher than 0.13%, it is detrimental to surface finish and may cause flaking. Leaded steels which are soft and malleable would appear suitable but this is not so, for lead has the same effect as sulphur.

In general the hardness should not exceed Rockwell 28. As a result of the cold working the hardness of the material is increased by Rockwell 8 and a surface finish of 10 to 15 μin is normally obtained.

THE RACK PROCESS

The rack process (Michigan) is shown in Figure 1.13(b), the two rolling racks being tapered so that the initial depth of engagement is small and the penetration increases as the racks are traversed and the metal flows. The work is completed in a single stroke in the one direction in some 3 to 5 s. The racks are made of high-speed steel with a fine grain structure. A low-viscosity oil is applied with extreme pressure additives, and as high beam loads are applied to the teeth it is only suitable for high pressure angles, 25° or more, and for stub teeth. Spur and helical gears can be produced by the process; for helical gears the rack dies are made to the appropriate angle.

THE CIRCULAR DIE PROCESS

The forming dies are circular, the component being placed between them so that the face width can be produced simultaneously. The diameter to be rolled is not limited by rack length. For producing teeth on bars the Maag system, in Figure 1.13(c) (Landis) two dies are used opposite and contra-rotating so that the work can rotate and be fed between them. The dies are actually worms and the process is continuous indexing and generating. While the dies rotate around the axes they are oscillated in an elliptical path which moves the dies in and out of engagement and gives a rapid hammering effect inducing plastic flow in the material. Gears up to 5 in. (127 mm) are being produced and helical gears up to a helix angle of 45°.

GEAR TOOTH GRINDING AND SHAVING

For high-precision gears, grinding of the teeth after hardening is essential and various forms of generating systems are available. One of these comparable with gear hobbing, is the Reishauer system (Vaughan Associates Ltd) which uses an abrasive worm as shown in Figure 1.14 in mesh with a spur wheel for grinding the teeth; Figure 1.15 shows the operation on a helical gear. For the latter operation it is merely necessary to swivel the machine saddle to the angle required to suit the helix of the gear. The surface of the grinding worm has a single- or multiple-start rack profile which corresponds to the geometry of the gear being ground. During the grinding process, the rotating worm meshes continuously with the teeth of the workpiece and thereby generates the involute tooth profile.

At the same time, the workpiece is moved along its axis in several

Figure 1.14
The Reishauer system of grinding gear teeth (by courtesy of Vaughan Associates Ltd)

strokes. At the upper and lower reversal points, the grinding worm is automatically fed in step by step until the desired base tangent length is attained. Cutting is done in both directions, the point of contact between worm and work being cooled by oil. The grinding worms have a maximum

Figure 1.15
An illustration showing the grinding of helical teeth

diameter of 350 mm and a width of 84 mm, so that a high tool life is obtained. For example, up to 50 gears can be ground between two dressing operations, ie up to 15 000 gears can be ground with a single grinding worm.

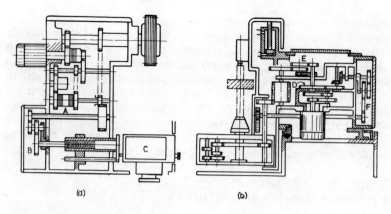

(a)

(b)

Figure 1.16
The design features of the Reishauer gear grinding machine

DESIGN FEATURES

Figure 1.16 shows the design features of a Reishauer machine, from which it will be seen that there is no mechanical connection between the grinding worm drive (a) and the work spindle (b). They are driven by two independent synchronous reaction motors running in absolute uniformity, while the work drive is equipped with an adjustable brake pump to prevent fluctuation of loads. The horizontal grinding slide executes the in-feed movement, the driving motor driving either the grinding spindle alone or the spindle and dressing drive for profiling the worm.

The clutch for the dressing slide is shown at A, while the change gears for dressing are indicated at B, these driving the lead screw of the dressing slide C. For the work drive the motor is attached to the swivelling support of the work slide so that the motor axis is kept parallel to the axis of the workpiece. When grinding helical gears, the wheel blank is given an additional rotary movement by the differential gear D. The change wheels for the number of teeth are shown at E, and those for the helix lead at F.

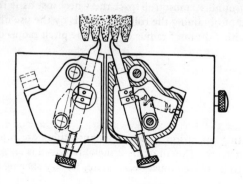

Figure 1.17
The method of truing the grinding wheel by diamond dressers

DIAMOND TRUING

Figure 1.17 shows the universal device which dresses both flanks at the same time. Two independently adjustable heads guide the diamond tools which are moved along the flanks, and profiles with pressure angles of between 14° 30′ and 30° can be dressed. The grinding worms can also be formed by crushing, the crusher being pressed hydraulically against the slowly rotating grinding wheel. The crusher is moved at a speed predetermined by the pitch and revolutions of the grinding worm which drives it. Small tooth profiles of 1.25 mm (a diametral pitch of 20) can be profiled directly into the solid grinding worm.

PRODUCTION EXAMPLE

A production example is given in Table 1.2

Table 1.2
A production example

Helical gear	30°
Number of teeth	56
Module	3 (diametral pitch of 9)
Pressure angle	20°
Face width	30 mm
Material	16 Mn—5 Cr
Hardness	Rockwell 62C
Grinding time	14.5 min
Production time	8 hr
Number of gears	28

ALTERNATIVE GEAR GRINDING METHODS

The formed grinding wheel process is used where the flanks of successive teeth are ground in turn as the formed wheel passes between them. The blank is then indexed at the end of each return stroke. It is necessary to compensate rapidly for wheel wear. An alternative generating system uses the concave side of an abrasive wheel to represent the side of a rack tooth. As the gear is being ground it must roll past the wheel just as if rolling along a rack. A method of obtaining the rolling action is by the use of steel tapes in conjunction with a drum of radius equal to the pitch radius of the gear.

GEAR SHAVING

PRINCIPLE

Gear shaving is a cutting operation in which minute chips are removed from the flanks of gear teeth. The amount of metal removed is extremely small so that the gear to be shaved must be accurate. Rotary shaving tools,

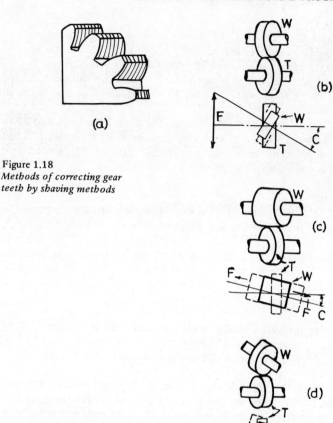

Figure 1.18
*Methods of correcting gear
teeth by shaving methods*

T-TOOL W-WORK
F-FEED C-CROSSED AXIS

as in Figure 1.18(a) are similar to a hardened and ground gear, of spur or helical form, with a number of serrations running from the crests to the roots of the teeth. These cutting serrations run into clearance grooves at the root of the spaces.

In use the shaving tool is meshed with the gear to be shaved so that they rotate in tight mesh with the tool driving the work. The tool and gear are of different helix angles, this resulting in what is known as the 'crossed axes angle'. For a helical gear the tool and gear must be of opposite hand of helix. The crossed axes between tool and work provide the necessary sliding velocity across the teeth to give the cutting action. The most efficient angle has been found to be between $10°$ and $15°$.

ADVANTAGES

Shaving is an efficient means of reducing errors left by the initial machining operation and improves concentricity, tooth profile and surface finish. It is used extensively for automobile gears making a precision gear from a good commercial one, but it is not generally used for the main driving gears of machine tools where, owing to the hardness of the material, grinding is used.

Shaving is used for gears with a diametral pitch between 3 and 80 (a module of between 9 and 0.3) with a hardness between Rockwell 25C and 30C. Helical gears up to a maximum helix angle of $55°$ are suitable together with double helical gears with a central gap.

SHAVING METHODS

There are several methods in use, but the method used to shave a particular component will determine the face width of the tool.

Underpass

Figure 1.18(b) shows this method, in which the work moves tangentially to the tool. This is the fastest of the three methods and maximum tool life is obtained. It is suitable for gears up to a face width of 1½ in. (38 mm) and for shoulder gears.

Transverse

The work is reciprocated axially, while the feed is radial (Figure 1.18(c)). The method enables low-cost narrow tools to be produced and is best suited for finishing gears with a face width of over 2¼ in. (57 mm).

Traverpass

Traverpass (Figure 1.18(d) is also known as diagonal shaving. This is a combination of the other two methods and is useful for shaving gears with a face width of from 1½ to 2¾ in. (from 38 to 70 mm) but is not very adaptable for shouldered gears.

Although the life of a shaving tool is comparatively long, shaving tools should not be continued in use when blunt because otherwise the effective life is considerably shortened. The tools should always be returned to the maker for resharpening.

CUTTING SPIRAL BEVEL GEARS

Whereas straight-tooth bevel gears find wide application in general power transmission, the advantages of spiral bevel drives include silent running owing to the simultaneous engagement of several teeth, high imunity to misalignment or deflection under load and greater capacity to transmit and redirect dynamic forces.

OERLIKON SPIROMATIC BEVEL GEAR CUTTER

Figure 1.19 shows the British Oerlikon number 3 machine (which is one of a range) with a capacity for cutting crown wheels with a maximum diameter of 640 mm at a mean spiral angle of $30°$ and a module range of 2.65 to 13. The maximum transmission ratio is 1 to 9.5.

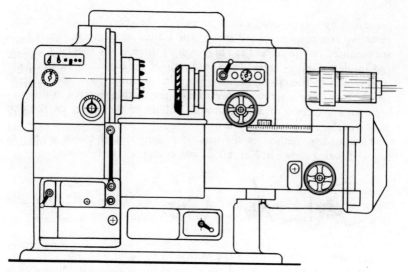

Figure 1.19
A diagram of a bevel gear cutting machine
(by courtesy of British Oerlikon Ltd, Birmingham)

The machine bed houses the main motor and the drive to three sets of change gears mounted at the front of the bed, ie a set of roll motion gears, the drive change gears for setting the cutter head speed and with it the blade cutting speed, and the roll feed change gears. Above the gear trains lies the roll cradle drive head, while the opposite side of the bed is designed as a circle segment to allow the swivel table with the workpiece spindle head to swing around the machine centre axis. A hydraulic unit controls the rapid traverse and feed of the drive head during its working phases, its operation being similar to a copying system. The roll cradle in the drive head carries the tilting cutter spindle powered by the motor and gearing, and the rotation of the cradle produces the rolling motion which is a feature of the system to be described.

TOOTH GENERATION
With the Spiromatic system the longitudinal crown on the tooth is

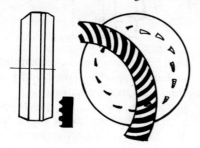

Figure 1.20
The setting of blades on the cutter head for spiral bevel teeth

generated by a compound rotary motion of the gear blank and the cutter revolving at a pre-determined speed. The blades, as shown in Figure 1.20, are mounted on the cutter head in groups of three, ie the roughing, outside and inside finishing blades. The speed ratio of the cutter to the workpiece corresponds to the ratio of blade groups in the cutter to number of teeth on the gear. The lengthwise curvature of the teeth generated in this way represents a section of an epicycloid. Rolling of the cutter on the gear produces the involute tooth profile.

The cutting method is the same for pinions and crown gears, the sequence being shown in Figure 1.21 and comprises the following:

Figure 1.21
The sequence for cutting spiral bevel teeth

(1) The cutter approaches the workpiece by rapid traverse.

(2) By plunge cutting the tool enters the blank, without a rolling motion and cuts the teeth to full depth by the continuous indexing method.

(3) The roll cradle starts to rotate. Both finishing blades, ie the inside and outside blades of one group now roll along the gear tooth flank and generate the tooth profile. Finally, the headstock and roll cradle return to the starting position at a rapid traverse.

SPIROFLEX SYSTEM

The Spiroflex system is another Oerlikon method in which the spindle can tilt in all directions, as shown in Figure 1.22. A feature is the low cost of

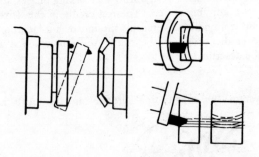

Figure 1.22
The Spiroflex system used on the Oerlikon machine

the blades which are simple to make from a rectangular bar and are mounted in the cutter head on a ring which sets them to height only, for no radius setting is required. In this system the crown along the tooth length is generated by inclining the cutter head spindle and so cutting a little deeper at both ends than at the load-bearing centre. Quiet running is obtained by the feature that the tooth contact lines and the cutting lines do not coincide but cross each other, owing to the continuous indexing.

INSPECTION OF GEAR TEETH

The chordal thickness of a gear tooth can be found by a gear tooth caliper provided with sliding verniers set at 90°. Adjustment can be made to the jaws that are integral with the vernier slides so that the thickness of a gear tooth can be gauged at any pre-determined distance below the tip of the tooth. It is usual to measure chordal thickness between those points of a tooth that lie on the pitch circles and the radial distance from those points to the tip of the teeth as shown in Figure 1.23.

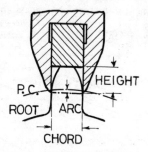

Figure 1.23
A gear tooth caliper for measuring the chordal thickness PC is pitch circle

The use of the tangent comparator has increased because, in relation to the gear tooth caliper, the following considerations must be made.

(1) Measurements are made without using the tip circumference as a location point, thus absolving the user from making allowance for ecentricity between tip and pitch circle diameters.

(2) One-dimensional setting, that of base tangent length only, is required.

(3) As the anvils are at a tangent to the flanks, a definite 'feel' is obtained, giving closer accuracy in measurement.

In the use of this instrument (Figure 1.24) it is necessary to know the

Figure 1.24
A method of finding the length of the base tangent

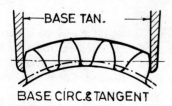

length of the base tangent over a given number of teeth, together with specified conditions of backlash. Nominal base tangent lengths for spur and helical gears can be obtained from the formula

$$d_0 \cos \sigma_0 \left(\frac{g \sec \sigma}{d} + \operatorname{inv} \psi_t + \frac{n\pi}{t} \right)$$

where d_0 is the base diameter and is equal to $d\psi_t$, d is the diameter at which the normal tooth thickness is g, σ is the helix angle at a diameter d, t is the number of teeth in the gear, n is the number of tooth spaces between the teeth to be gauged, inv ψ_t = tan $\psi_t - \psi_t$ (rad). For spur gears the formula cos σ_0 = 1 and sec σ = 1 should be employed.

GEAR TESTING MACHINES

It will be appreciated that hand methods of checking are not suitable where large numbers of gears are manufactured and that a more rapid system is required. The range of gear testers produced by J Parkinson & Son Ltd, Shipley, includes models for testing spur, helical, worm and bevel gears. Figure 1.25 shows a machine for testing spur gears in pairs and,

Figure 1.25
A gear testing machine (by courtesy of J Parkinson & Son Ltd, Shipley)

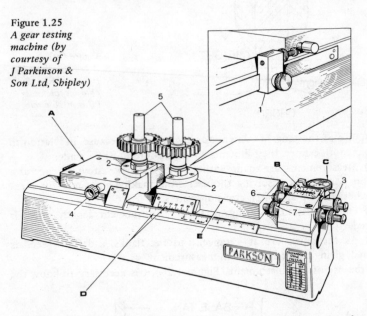

though not essential, one of these may be a master gear or one known to be correct. The gears are mounted on arbors 5 which are located on separate slides A and E, the first being adjustable for centre distance and then clamped by knob 4. The floating slide is supported on needle roller bearings and is spring loaded by nuts 6.

The regulating nuts 7 act against the spring load so that the distance between the gears can be adjusted to a definite amount, while an adjustable stop 1 allows the adjustable slide to be moved away and brought back to its original position. For testing, the gears are rotated slowly, and a pair of theoretical gears, assumed perfect, thus revolving would not show any centre distance variation on the dial indicator C, but any defects of eccentricity or tooth form are shown by the pointer on the dial. A scale and

vernier D is available for setting an accurate centre distance, but an alternative method is to mount standard discs of the correct pitch circle diameter
on the arbors with the peripheries touching.

Individual gears can be tested for effective diameter, depth of tooth and
eccentricity by mounting them on one arbor with a pin or roller in a tooth
space. The gear is rotated so that the pin contacts the arbor on the other
slide. Repeating this for other teeth, the indicator shows any variations
and also checks the effective diameter. To check backlash, the floating
slide is clamped, and a calculation is made by noting on the indicator the
distance the slide moves when the adjusting nuts 7 are unscrewed, allowing
the teeth of the gears to mesh close without backlash and applying this
movement into the formula

$$X \times \tan (PA) \times 2 = Y$$

Conversely

$$Y = \frac{Y}{\tan (PA) \times 2X}$$

X is the movement given on the dial, Y is the amount of backlash and PA
is the pressure angle.

Where for helical gears, since the movement is affected by the helix
angle, the formula is

$$\text{transverse PA} = \frac{\text{tangent normal to PA}}{\cos (\text{helix angle})}$$

$$X \times \text{tangent traverse PA} \times 2 = Y$$

The same principle of inspection is used for bevel and worm gears, but
more complicated mountings are required for the adjustable slide. If
graphic readings are required, the gauge heads of the recording equipment
can be readily mounted to work in conjunction with the testers. If power
rotation of the gears under inspection is required, a motor drive can be
provided to give a variable arbor speed range of 1½ to 15 rev/min in either
direction, the power driven arbor in the adjustable carriage rotating on precision balls.

MAAG GEAR TESTING MACHINES
An example of a typical machine is shown in Figure 1.26. For tooth
profile inspection the workpiece is attached to the rotating mandrel which
carries a disc A with a periphery of the same diameter as the base circle of
the gear to be inspected. The disc is pressed with the carriage against the
straight edge B. As the cross slide is moved in the direction indicated by
the arrow, the base circle disc, in spring-loaded contact with the straight
edge, rolls without slip on the latter. During this generating action, the
edge of the feeler C attached to the vertical slide is kept by spring pressure
in contact with the tooth flank of the gear. The feeler edge is exactly

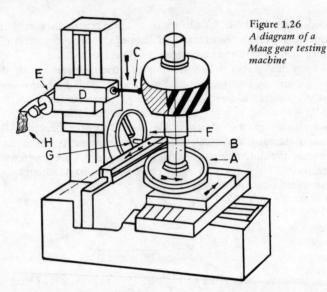

Figure 1.26
*A diagram of a
Maag gear testing
machine*

above the edge of the straight edge and, if the flank profile is a true involute, the feeler remains stationary but, if the profile deviates from the involute, the feeler is deflected to one side. The deflections are scaled up in the recording head D and are transmitted to the pen E which produces the profile diagram.

For tooth alignment and helix angle inspection, the slotted guide disc F on the vertical slide is set via an optical scale to the base helix angle of the gear. The cross slide is fixed in the position at which the feeler contacts the tooth flank at the desired profile point. As the vertical slide is moved up or down, and with it the guide disc, the sliding block G thrusts the straight edge in the direction indicated by the arrow.

Through the spring-loaded contact between the straight edge and the base circle disc, the latter is rotated along with the gear. Thus the feeler contacting the tooth flank receives a relative movement along a tooth trace of the theoretical correct helix angle. Any discrepancy between the actual helix angle and the optically set base helix angle causes a deflection of the feeler which is scaled up and registered on the chart H as a deviation of the straight line. The electro-mechanical recording device has a switch to reverse the direction of the feeler pressure to permit inspection of right and left tooth flanks. The error curve is produced by the pen on metallized paper with square co-ordinates, and the chart feed is transmitted true to scale from the drive elements of the two slides. Synchro-generators are fitted to the lead screws of these slides, which emit electric impulses according to the rotation of the screws. The impulses are amplified electronically and are fed to a stepping motor which drives the chart drum.

Error point location

Since the scale of the diagram varies with the point along the profile, a

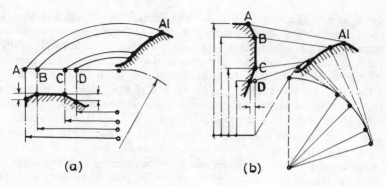

Figure 1.27
A method of using error point location

point on the diagram cannot be projected onto the tooth profile by linear proportion. On the standard machine the length of the diagram is proportional to the peripheral segment through which the base circle is rolled (Figure 1.27(a)). An alternative is to locate points along the profile in degrees of roll (Figure 1.27(b)). Generating paths to an accuracy of 0.1 mm are read on the cross slide vernier scale.

The operating principle of another Maag machine can be seen from Figure 1.28. The gear to be tested is clamped with its axis coincident with

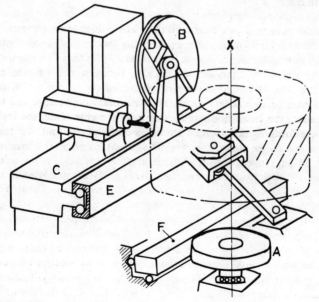

Figure 1.28
The operating principle of an alternative Maag gear testing machine

the x axis of the base cylinder A. The carriage with the slide assembly, cross slide, vertical slide, guide disc B and the feeler head can be moved towards and away from the workpiece. It is set by an optical scale so that the feeler tip lies in the plane tangential to the base cylinder of the gear.

During the profile test the cross slide C with the measuring stand and feeler are moved in a straight transverse line. Sliding block D engaging in the guide disc provides the connection between the cross slide and the helix bar E. From this bar which traverses the same length as the cross slide, the motion is transmitted to the straight edge F via a pivoted extending linkage, the straight edge moving a proportional distance. Its frictional contact turns the base cylinder and gear through the required angle of rotation. The locus of the feeler relative to the gear takes the form of an involute.

Helix locus

The guide disc is set by an optical scale to the base helix angle and the carriage is again set to the base circle of the gear. The cross slide is adjusted and set in the position in which the feeler touches the flank of the tooth at the desired point on its profile. During measurement, the vertical slide and guide disc are moved vertically. Since the height of the sliding block does not change, it is displaced transversely by the guide slot a certain distance, carrying with it the helix bar. Via the link bar, straight edge and base cylinder, a rotation is transmitted to the gear under test, corresponding to the rotary movement of the helix locus. The feeler moves up and down along a tooth trace.

Tooth profile inspection (involute)

The distance between the work axis and the feeler, which must correspond exactly to the base circle radius of the gear being checked, is adjusted with the aid of an optical setting device. A built-in base cylinder and a magnetic straight edge form the non-slip connection between the straight-line motion of the cross slide and the rotary motion of the workpiece. Whatever the position of the carriage, ie whatever the base circle setting, a lever system correlates the cross slide motion and work rotation automatically.

Whilst measuring is in progress, the workpiece rotates and the cross slide carries the feeler along the rectilinear component of the involute generating motion, tangential to the base circle.

2

HONING
AND SUPERFINISHING
PROCESSES

THE HONING PROCESS

Honing is fundamentally a wet-cutting process comprising the mechanical application of bonded abrasive, or diamond, to an internal work surface. It accomplishes four purposes

(1) It removes metal.
(2) It generates size.
(3) It maintains accuracy of roundness and straightness.
(4) It produces a high surface finish.

The honing tool consists of a holder carrying up to six abrasive sticks which fit in slots; it has provision by taper bushes, or other means, to force the sticks outwards against the side of the bore to be honed. The cutting pressure can be controlled either by hand or automatically by adjusting the taper bushes and so causing radial movement of the sticks.

The tool is connected to the machine spindle by a flexible coupling and is thus not affected by any inaccuracy in the running of the spindle. The sticks are arranged to give an even pressure distribution, the hone floating in the bore. Low unit heat and pressure is a feature of the process, which gives a uniformly controlled cutting action obtained by a multi-directional travel path of the sticks, in various combinations of rotation and traverse.

CROSS HATCH PATTERNS

Figure 2.1 shows a harmonic travel path, in which for simplicity the ratio is slightly more than two revolutions to one reciprocating movement. The path on the single point on the hone would be as shown in the lower right-hand corner. The crossing of such paths on the forward and reverse strokes forms the symmetrical cross hatch surface pattern which is characteristic of the process. Accuracy for roundness is maintained by the rotary motion of the hone, and accuracy for straightness by the reciprocating or reversing traverse. The degree of surface smoothness is controlled by the adjustment of the cutting action, usually by reducing the cutting pressure or by choice of abrasive.

Honing uses a greater number of cutting points than are found in almost any abrading process at one time. For example, in a bore 3 in.

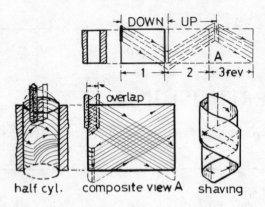

Figure 2.1
A diagram showing the travel path of cutting grains when honing

(76 mm) in diameter and 8 in. (203 mm) long, 150-grit stones would have a total area of 7.5 in.^{2} (48 cm^2), in which there would be about 98 000 simultaneous shearing contacts. A corresponding internal grinding wheel using a 46-grit wheel would have about 0.05 in.^{2} (1.5 cm^2) of total contacting area of the work with only about 48 simultaneously stock-removing contacts at one time.

The total normal pressure, conventionally used in honing, results in unit pressures ranging from about 55 to 75 lbf/in.^{2} (from 3.8 to 5.3 kgf/cm^2) but may be lower for fine honing. These pressures distributed over thousands of cutting points result in a pressure per grain which ranges from 0.003 to 0.015 ozf (from 0.08 to 4 gf) for finish honing and from 0.02 to 0.34 ozf (from 0.56 to 9 gf) in rough honing. These extremely low pressures are exerted at the lowest speeds used in any mechanical abrasive methods.

HONING MACHINES
Honing machines are made in two types, either vertical or horizontal in construction; horizontal machines are more suitable for honing long bores, and small horizontal machines have been prominent for a long time. The spindle rotation is generally by mechanical means or hydraulic motor rotary drive, while the spindle reciprocation is usually hydraulically operated. The speed ratio of the two motions affects the work finish and may be varied throughout the operation or for different materials. For cast iron the speeds range from 200 to 500 ft/min (from 60 to 150 m/min) for rotation with 50 to 70 ft/min (from 15 to 21 m/min) for reciprocation. The corresponding speeds for steel are from 150 to 200 ft/min (from 45 to 60 m/min) and 40 ft/min (12 m/min) for reciprocation.

Figure 2.2
A vertical honing machine (by courtesy of Delapena Honing Equipment)

Figure 2.2 shows a vertical honing machine (Delapena Honing Equipment). This is the standard single-spindle machine with provision for simple work fixtures, but the machine is available with two spindles and with rotary indexing or 'cross-loading fixtures'. This type of fixture can have stations at either side for locating components for the separate honing of two bores. A loading pattern is employed whereby, while two components are being honed at the rear, one of the front two is moved to the other station, from whence a finished component is removed. A new component is then introduced into the empty station. This rotation of work ensures that one complete component is obtained each cycle.

The machine includes a 'drop pressure hydraulic system' whereby, after the actual honing stage, the machine can be set to continue the operation for a pre-determined period at a reduced stone pressure merely to polish the workpiece bore without removing any detectable amount of stock.

HORIZONTAL HONING MACHINES

Delapena horizontal honing machines have been developed for tube honing; some of these are alloy steel forgings for turbine rotors ranging up to 5 ft (1.5 m) in diameter, 60 ft (18 m) long and with a mass as large as 80 tonnes. Machines such as that shown in figure 2.3 may have a stroke of 30 ft (9 m);

Figure 2.3
A horizontal honing machine for long tubes

the work is reversed if a longer bore requires to be honed. The travelling carriage that holds the hone spindle runs on two steel bars attached to the bed of the machine. It is driven along these round bars by a duplex chain that connects two sprocket wheels at opposite ends of the bed, one of the sprockets being driven by a hydraulic motor. The speed adjustment of the carriage is by a variable delivery pump that supplies the hydraulic motor, the maximum speed being 80 ft/min (24 m/min).

The chain also drives three travelling steadies that support the hone drive shaft. Two rollers on each steady support the weight of the shaft while the third one prevents any lift due to whirling. As the carriage moves the rotating hone shaft into the bore of the work, the steadies follow in step until they reach the end position near the work. They are then, each in turn, brought to rest. When reversal of the carriage takes place, the steadies are again moved in sequence. The rotational speed of the hone is steplessly variable between 30 to 300 rev/min. Honing fluid is conveyed to the carriage and is fed inside the drive shaft to the hone head.

HYDRAULIC CIRCUITS

Figure 2.4 shows a circuit developed by Vickers, Sperry Rand Ltd and applied to Delapena honing machines. The servo pump B is connected in a closed circuit to a hydraulic motor C coupled to the reciprocating carrier of the machine. Only small forces are required to position the piston pump

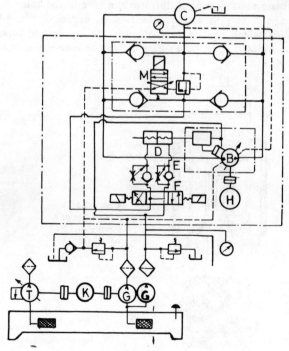

Figure 2.4
A hydraulic
circuit for honing
machines (by
courtesy of Vickers,
Sperry Rand Ltd)

swashplate, and this is done by a small cylinder D and a Flue-trol stack with flow and directional control modules at E and F. The double-vane pump G provides the boost supply as well as oil for actuating the stem servo cylinder of pump B driven by a 20 hp electric motor H, while pump G is driven by a 7½ hp motor K.

Automatic reciprocation takes place by energizing the solenoid valve M which closes the vent line for the relief valve F, thus raising pressure up to a maximum of 1,415 lbf/in.^{2} (100 kgf/cm^2). Automatic cycling takes place through limit switches which reverse the setting of valve F, thereby reversing the position of the pump actuating cylinder D. When the pump reverses from maximum delivery through zero, dynamic braking of the honing head mass is controlled by the inclination of the swashplate. Thus the pump is employed as a flow control valve which is connected mechanically to the electric motor running at a synchronous speed of 1,500 rev/min. The pressure-compensated pump T serves for auxiliary functions on the machine.

HONING TOOLS

Figure 2.5 shows a typical honing head, the body being a tube with two collars in which six slots are milled. Small slots are also machined from the bottom of these main slots extending through the bore of the tube. Six stone holders E fit in these slots and are adjustable by expansion plates D inserted in the slots which extend to the bore. The upper and lower expanding cones C are held apart by a spring but can be moved axially by

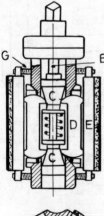

Figure 2.5
The design of a honing tool with six stones

a central control rod B. When the cones are screwed towards each other, they press upon projections of the expansion plates and push out the stone holders to an increased diameter. The stone holders are retained in contact with the expansion plates, and with the cones, by means of light tension spring rings G placed over the projections on each end of all the stone holders.

The stones are held in the holders with lead or light metal, and by means of a fixture these can be set in correct relation to the necessary cylindrical form, a useful accessory when assembling new stones.

Figure 2.6 shows the expansion of stones in a honing head on a vertical

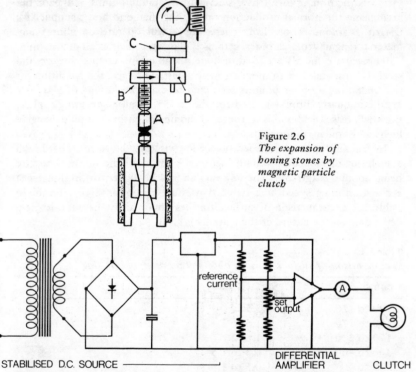

Figure 2.6
The expansion of honing stones by magnetic particle clutch

STABILISED D.C. SOURCE ————————————┘   DIFFERENTIAL AMPLIFIER   CLUTCH

machine made by Fuji Seiki of Japan. The stones have wedge-shaped inner faces operated by cones on bar A which has a threaded end, the nut B being turned to move the bar axially. The drive is taken from the motor through worm gearing, a magnetic particle clutch C and reduction gearing D.

As shown in the detail, by varying the value of the current supplied to the clutch, the amount of magnetism induced in the particles is changed, and the total torque which it will transmit is thus controlled. Thrusts exerted in expansion of the honing stones can be varied over a wide range and can be controlled very closely during an operation. The machine has a spindle driving motor of 5 hp which provides 1 hp for each 1 in. bore of the workpiece.

DIAMOND HONING

The diamond honing process is finding increasing use in industry owing to the wide range of applications which have been successfully changed to diamond honing with a saving in time and cost. Mass production rates can now be maintained with a geometrical tolerance in some cases as low as 0.00002 in. with a surface finish of the order of 5 to 7 μin. Such an application is the honing of nitrided sleeves of hydraulic pumps, 100 parts per hour being the normal production rates using diamond hones supplied by Diagrit Diamond Tools Ltd. These hones will operate on almost any material ranging from carbides, ceramics, glass and steels of all grades.

Three basic metals are used in hone manufacture, copper, bronze and steel. Each metal is then modified with additives to give the condition for the operation. Copper bond is restricted to diamond honing of glass and synthetic quartz. Bronze is used for about 75% of all engineering applications whereas steel bond is restricted to honing of chromium-plated engine liners and similar components.

In general, for roughing operations a low-viscosity lubricant is used with a high cutting speed to obtain high stock removal, the honing pressure being about 12 kgf/cm^2. The viscosity should be higher for finishing with a slower honing speed at a reduced pressure to obtain a fine finish. In Table 2.1 a comparison of surface finishes obtained by diamond honing, depending on the nature of the material is given.

Table 2.1
A comparison of surface finishes obtained by diamond honing

Grain size	Surface finish (μin.)
100 to 120	3.5
150 to 170	1.5 to 3
170 to 200	1 to 2
300 to 400	0.8 to 1.5
20 to 40	0.5 to 1

DELAPENA DIAMOND HONE

A Delapena diamond hone is shown in figure 2.7 and it includes an expansion wedge assembly to give an accurate relationship between the advance of the wedge and stone movement. This is effected by tne push rod A, giving a motion to operate a micro-switch B when a specified honing diameter has been reached. Since diamond hones are employed the rate of wear is slow, and adjustment of the tripping screw to compensate for this is a simple operation. The diamond hones are indicated at C.

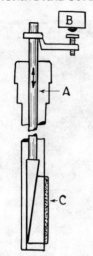

Figure 2.7
*The expansion
wedge assembly
of a honing tool*

As an indication of production times, 0.004 in. (0.1 mm) of metal is removed from the bores 110 mm in diameter of cast iron liners with an average cycle time of 60 s.

SUPERFINISHING

Turned or ground surfaces have certain defects caused by the cutting action which in grinding produces fragmented metal and often annealing of the outer layer by the pronounced, although momentary, heat of the grinding operation. Temperatures in excess of 2000°F are often developed in spite of a generous supply of coolant. To remove these surface defects, the process of superfinishing was introduced. Examples of these defective surfaces are indicated in *Manufacturing Technology*, Volume I, *Basic Machines and Processes*, Chapter 9. Relatively coarse bonded abrasives are used with an action which comprises a short scrubbing motion, light pressure, slow abrasive cutting speed and a fluid acting as a lubricant instead of a coolant. The surface produced is equalled in quality only by the finest lapping but with the basic advantage that the surface is smoothed in a few seconds. The abrasive blocks are of such a surface area that, once any grit has done its share of cutting, it is washed away and does not remain to produce scratches. The blocks are made in sizes to suit the work and may be rectangular, curved or cupped.

The abrasives used are aluminium oxide for materials of high tensile strength such as carbon, alloy and high-speed steel and materials of considerable hardness. Silicon carbide abrasives are preferable for low-tensile-strengh materials such as cast iron, aluminium, brass and materials such as brake linings. Bonded diamond dust is used for finishing cemented carbide tools.

The number of stones used depends upon the size of the work; up to 3 in. (76 mm) in diameter a single stone is satisfactory, whereas for large

Figure 2.8
A machine for superfinishing flat surfaces

rolls four or six stones may be mounted in a single holder. Flat work as indicated in Figure 2.8 allows the advantages of a multi-directional path, because the abrasives working in this plane are not limited as they are in cylindrical work. Self-contained heads, similar to that shown in Figure 2.8, are available for fitting on a lathe rest.

For superfinishing round work, saddle-type stones, as shown in Figure 2.9, are used; these may be, for example, oscillated 1/16 in. (1.5 mm) 500 rev/min, making 1,000 strokes/min. The mechanism oscillating the stone may itself be oscillated ¼ in. (6.3 mm) at 100 rev/min, making a motion of 200 strokes of ¼ in. Because of the combination of the longer

Figure 2.9
*A saddle-type
stone for super-
finishing round
work*

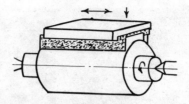

slower stroke and the shorter faster stroke, the abrasive grain must follow
a path of ever-changing direction at varying speeds. The traversing feed
which is distinct from the scrubbing action, is somewhat faster than that
employed for turning, about 3/16 in./rev (5 mm/rev) being usual.

Light thread grinding oil is a good lubricant, but paraffin compounds
can be used. The pressure in superfinishing is low compared with grinding,
being about 20 lbf/in.^{2} (1.4 kgf/cm^2) of the stone area, and this low
pressure allows a surface finish of 1 to 4 μin. to be obtained.

LAPPING AND LAP HONING

Mention has been made in *Manufacturing Technology*, Volume 1, *Basic
Machines and Processes* Chapter 6, of the basic process of lapping plane
surfaces. Hand lapping is occasionally used, but more frequently a machine
is used to give consistent results in a short time. The process of surface
finishing using a plate with applied abrasive is known as lapping whilst the
use of a grinding wheel for the process is known as lap honing. A machine
suitable for both these processes is shown in Figure 2.10; it carries two
grinding wheels or two lapping plates, the work is placed between the two
wheels in a suitable holder and the upper wheel is lowered hydraulically to
apply cutting pressure. A rotating eccentric pin in the centre of the wheels
is used to oscillate the work across the face of the wheels. The eccentric pin
rotates at a speed unrelated to the speed of the wheels so that any parts
being machined follow a slightly different path across the grinding wheels
at each revolution of the wheels, thus making wheel wear as even as
possible. A diamond dressing device is fitted to the machine and is seen on
the right side of the illustration; the arrangement which provides even
wheel wear ensures that dressing takes place as infrequently as possible and
maintains geometrical accuracy of the parts being machined. The machine
is made by Armstrongs (Engineers) Ltd, carries wheels 24 in. in diameter
and is one of a range with wheels up to 36 in.

Small parts require a different technique which is illustrated in Figure
2.11; the workpieces are tungsten carbide inserts for turning tools and are
loosely held in gear-shaped sheet metal fixtures; they have freedom to
move slightly in the pockets. Rotation of the inner plate-carrying pins
which act as gear teeth causes epicyclic rotation of each work-holder gear.
Thus, whilst the grinding wheels rotate, the work-holder gears carry the
parts around the machine guided by the outer fixed pins and carry each
part across the full width of the wheels. The eccentric pin referred to in

Figure 2.10
An Armstrong lapping machine equipped with two abrasive wheels for lap honing

the previous paragraph for lapping large parts can be seen near to the centre of the inner plate. The upper grinding wheel is removed for the purposes of this illustration.

Figure 2.11
Planetary work holding discs for Armstrong lapping machine

3

BROACHING MACHINES AND OPERATIONS

BROACHING PROCESS

The broaching process consists of the removal of metal by a tool with a series of teeth which increase progressively in height from the starting end, so that each tooth takes a light cut and, after one travel of the broach through or over the work, the required shape is produced. Unlike most machine tool operations, the feed is not applied by the machine but by the increase in height of each cutting tooth.

Internal broaching is used for forming round or square holes, holes with keyways, grooves or slots of all sections straight or helical, and internal gear teeth. For broaching multiple splines concentric with a bore, the bored hole may be left undersized and a combination round and spline broach may be used, as in Figure 3.1. Alternate spline and round cutting teeth finish the splines and hole at the same pass.

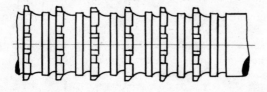

Figure 3.1
A combination round and spline broach

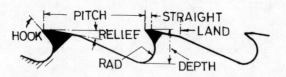

Figure 3.2
The elements of design for broach teeth

The depth of cut per tooth rarely exceeds 0.07 mm, and the heavy roughing cuts are taken at the start of the traverse and the light cuts at the end to ensure a good smooth finish. The cutting speed is constant throughout the stroke and varies with the material, but hardness is the deciding factor. The chief elements of tooth design are shown in Figure 3.2; the front rake or hook angle is governed by the material but varies from $6°$ for cast iron to $20°$ for soft steel. Aluminium requires $10°$ or more, and hard brass from $5°$ positive to $5°$ negative.

Pitch is based upon the length of the component, the nature of the material and the size of chip removed, an approximation being based upon the formula

$$\text{pitch} = 0.35 \times (\text{length of cut})^{1/2}$$

The straight land should be as short as possible, bearing in mind that re-grinding takes place on this section. The length is generally 0.01 to 0.02 in. (0.25 to 0.5 mm) on roughing and 0.005 in. (0.13 mm) on finishing teeth. Some slight backing off of the straight land is necessary to prevent it wearing to a negative angle, the amount being ½ to $2°$. The depth of tooth is determined by the pitch, but the radius is most important in that the chip should curl and should not pack tight in the space, for this may cause tooth breakage.

CHIP BREAKERS
Broaches are provided with chip breakers, for without these on a round broach, for example, it would be difficult to remove a full circular chip of considerable strength between each tooth. The correct design is obtained by ensuring that the chips do not come together at the bottom of the tooth. The breaker must be deep enough to be below the surface of the tooth ahead.

High-speed steel is commonly used in broach manufacture, a typical analysis being as follows: carbon, 0.69 to 0.72%; tungsten, 18%; chromium, 4%; vanadium, 1%. Tools tipped with cemented carbide are often used for surface broaching, but in fact any good-quality steel of the case-hardening type can be used. The heat treatment of hardening and refining should be carried out with the broach held vertically.

BROACHING MACHINES
Both horizontal and vertical machines are used, with hydraulic operation being employed for operating the drawhead on horizontal machines and for operating the saddles of vertical machines. As the return stroke of the broach is non-cutting, a high speed is employed to save the idle time. One method is to employ the oil against a large piston area for the cutting stroke and then to use it against a small area to increase the speed for the return. Figure 3.3 shows the operation using a hollow draw rod.

During the cutting stroke in the direction shown by the arrow, oil from the pump is delivered to port L. A proportion of the oil from the right-hand end of the piston passes to the tank M, as determined by the setting of the adjustable needle valve of the transfer valve N. The remainder passes

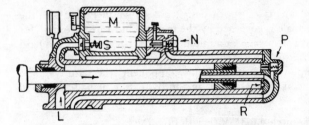

Figure 3.3
An example of a hydraulic horizontal broaching machine

through the automatic transfer valve P and is directed to the left-hand side of the piston without passing through the pump, by way of an auxiliary valve (not shown). For the return stroke, the latter valve is set automatically, and oil from the pump then passes to the tubular piston rod. As the piston is moved to the left, oil is drawn from the tank M through the transfer valve N to fill the right-hand end of the cylinder, the oil passing through the ports as at R in the end cover. The valve S remains closed during operation of the machine but is set to allow a small flow of oil from the tank M to the cylinder when the machine is idle, to ensure that the cylinder is always full of oil. With this system the ratio of cut to return speed may be as much as 1 to 4.

VERTICAL BROACHING MACHINES
An example from the range of The Cincinnati Milacron broaching machines is shown in Figure 3.4. The construction is of pre-fabricated steel with a range of operation which has a slide force of 3 to 40 tons-force and a slide travel of 0.75 to 2.4 m. On the left can be seen the Oilgear hydraulic pump unit which supplies power to the tool slide and to the shuttle table, or alternatively the table may be rotary or fixed. All speeds are steplessly variable. There are two push button stations positioned well away from the fixture, and these contribute to the operator's safety by demanding that he uses both hands to initiate the machine cycle. This can be selected for hand, semi-automatic or full automatic work cycle.

BROACHING FIXTURES
Details of broaches for connecting rods and caps are discussed later, but a Cincinnati arrangement is shown in Figure 3.5. The broaches for the half-bores and flat surfaces can be seen on a machine equipped with two identical automatic clamping fixtures, one for each station; both fixtures carry one rod and one cap. Both fixtures are loaded simultaneously and are automatically clamped. The use of progressive broaches completes the first stages of the broach cycle, slab-type broaches producing the serrations. The production time is 50 to 85 complete rods and caps at 85% efficiency, depending on the size of the units, in this case for large engines.

Figure 3.4
A vertical broaching machine of fabricated construction
(by courtesy of Cincinnati Milacron Ltd)

BROACH HOLDERS

There are several methods of connecting a broach to the pulling head of a horizontal broaching machine, the simplest of these being by a threaded connection or by a cottered joint. The faceplate of a horizontal broaching machine may be circular or rectangular, and the workpiece may be located on a spigot fitting in the bore of the faceplate. There is a diversity of practice as regards the method of working holding, but for simple operations it

Figure 3.5
An arrangement of the fixture and tools for broaching connecting rods

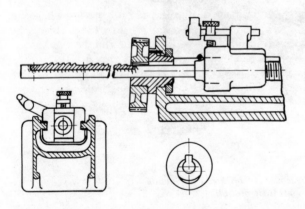

Figure 3.6
*A method of
broaching taper
keyways*

is merely necessary to place the broach through the opening in the work, to connect it to the drawhead and to pull the broach through the work which abuts the faceplate. In other cases, as in Figure 3.6, a simple adapter in the faceplate will enable taper keyways, for example, to be broached. In such a case the broach slides through the adapter and not the work, so that one broach can be used for varying bores and the number of broaches is dependent upon the width of the keyway required and not on the bores.

The broach may be built up as shown, a method that enables the tooth section to be replaced when required but ensures long life in that the teeth may be packed up after losing size through regrinding.

Some keyway broaches have a small shank, and in such cases a puller of the type shown in Figure 3.7 may be used. There is a screwed connection

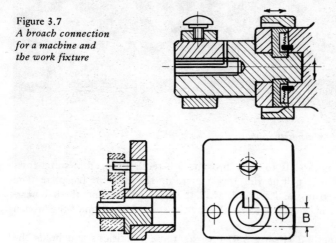

Figure 3.7
A broach connection for a machine and the work fixture

between the end of the broach and the puller, which has a loose half-nut held in position by a ring. With these removed, a broach can be placed in the puller without any screwing motion and can be locked there by the replacement of the half-nut and ring. There is also a quick-release device between the puller and the drawhead, this comprising an outer ring and spring-loaded plungers, which are either kept down in the driving position shown or, by sliding the ring, are freed from the puller by entering the large bore of the ring under spring pressure.

SPIRAL BROACHING

Helical grooves are often required in the bores of gears and in rifle and gun barrels. A typical broach with spiral teeth is shown in Figure 3.8. Such broaches are expensive to make and to resharpen and require careful handling. Nevertheless, the production of helical grooves without broaching is an almost impossible proposition.

Three methods are available, one of which is to attach a lead bar to the broach; this attachment has a helical groove similar to that required in the work. The lead bar and broach are fastened to the drawhead but are free to rotate, so that fixed pins contacting the groove cause the proper amount of rotation during the lengthwise travel of the broach.

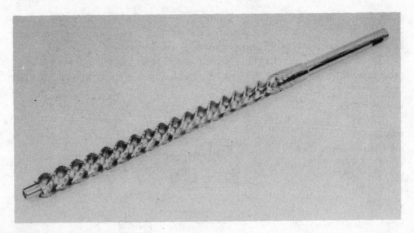

Figure 3.8
An example of broach with spiral teeth

Another method is to use a broach with teeth arranged in spiral rows and to provide a ball bearing thrust plate on the machine faceplate. The broach is fixed in the drawhead, while the work rests against the ball bearing support and is caused to revolve by the spiral formation of the broach teeth.

A third method (Figure 3.9) reverses these conditions and holds the

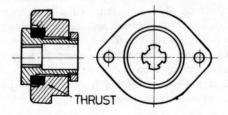

THRUST

Figure 3.9
*A lead bar used
for broaching
spiral grooves*

work rigidly while the broach has a ball bearing connection in the drawhead, so that now the helical teeth cause the broach to rotate as it traverses through the work. The broaches for rifling large guns are of the wafer type, and for a 40 mm gun 32 wafer cutters are required to produce the 16 grooves. The reason for this is that the helix increases in pitch from $4°$ at the breach to $6°$ at the muzzle, hence the reason for the number of cutters.

SURFACE BROACHING
Surface broaching follows conditions more usual in metal cutting than

does internal broaching, for the problems are simplified by the more open cut. In general, broaches have teeth in three sections, the long roughing section, the semi-finishing part and the finishing section which produces and holds the size. In machining fragile components the metal removal is progressive and can be arranged so that no part of the cut will cause distortion. However, large units such as cylinder blocks with heavy stock removal can be broached with equal facility.

As one of the problems in manufacturing is often that of determining whether surface broaching or milling will be the most suitable, the following considerations should be noted.

(1) The work must be strong enough to withstand the stresses set up by broaching.

(2) The faces to be broached must have all the elements parallel to the axis of the broach holder, and there must be no obstruction in the paths of these edges that trim these faces.

(3) The stock removal must be reasonably constant, and production quantities must be high in order to cover the cost of the broach.

(4) Broaching achieves good results because each tooth removes a fixed thickness of metal, producing light chips in finishing. To obtain the same result by milling, both roughing and finishing cuts are required.

(5) Broaching speeds are low, with each tooth striking the work once only and, as impact loads vary as the square of the velocity, wear is not as heavy on a broach as on a milling cutter, where speeds are about 10 times that of broaching, and each tooth strikes the work many times.

(6) Each tooth of a broach starts to cut immediately it touches the work, whereas each tooth of a milling cutter tends to rub the work surface before cutting commences, unless down-cut milling is employed.

(7) Undercuts can be machined with a broach at the same time as other surfaces, whereas milling requires other set-ups and fixtures.

(9) The cost of broaches is higher than that of milling cutters, and re-sharpening is a longer operation.

EXAMPLES OF SURFACE BROACHING

Connecting rods and caps form a common example and details of the broaches are shown in Figure 3.10(a) and Figure 3.10(b). Surface broaches are generally made in short lengths, about 250 mm long, because of the ease of replacement of an individual part, and also for ease of manufacture and hardening. Figure 3.10(a) shows that the bore is machined by the round broach A, and the two faces by the flat broaches B and C. Full-width broaching of the joint surfaces is assured by interlocking the teeth of all the broaches. The broaches are made in three sections so that, as the finishing part becomes worn, the other two sections can be moved forwards and a new section can be provided after regrinding. The teeth of the round broach are cut around the periphery so that, after one-half has become worn, it can be turned to use the other $180°$. The flat broaches are supported by the adjusting strips D and are clamped by vee slips E.

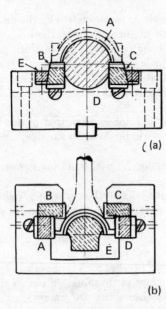

Figure 3.10
*Examples of surface
broaching connecting
rods and caps*

(a)

(b)

In the second operation (Figure 3.10(b)) four broaches A, B, C, D are in action together with broaches B and C cutting across the full width of their bottom teeth and for a short distance on their front teeth. The bottom teeth are interlocked with broaches A and D to ensure machining all surfaces along the full width. To ensure location of either cap or rod, the block E is shaped to that of the surfaces machined in the previous operation. The parts are machined on a double-ram machine from rough forgings with a stock removal of 5.5 mm. This requires a pull of 22 tonnes-force, and the production is 220 complete caps and rods/h. All broach assemblies are 1220 mm long with each section 356 mm in length.

THROW-AWAY TIP BROACHING
An impressive example of surface broaching using cemented carbide tips is that of machining the surfaces indicated in Figure 3.11 which represents a

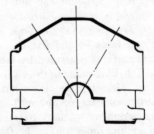

Figure 3.11
*A Ford cylinder
block broached
by throw-away
carbide tips*

diagram of the cast iron cylinder block for a vee-type Ford engine. All the cutting edges of the broaches are formed by cemented carbide inserts, and of these 2,228 are throwaway tips employed in the roughing stages. These tips are of a general-purpose grade for cast iron and are of a square form to provide eight cutting edges on each tip; in service, the average cutting life is 45,000 parts per cutting edge. For the ground inserts employed in taking finishing cuts, the average life before regrinding is necessary is 60,000 parts. Broaching is performed at a speed of 40 m/min.

BROACHING CRANK SHAFT BEARINGS

Surface broaching is not restricted to flat surfaces but can be applied to round components. Figure 3.12 shows the development in which the crank shaft revolves at 40 rev/min and the broach passes across the bearings at a feed of 7 m/min. The construction of the broach is such that not only the cylindrical surfaces but also the oil sling is machined.

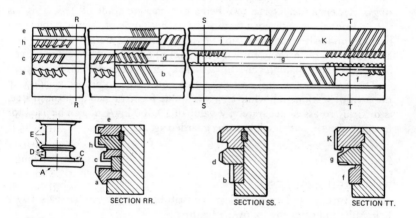

Figure 3.12
Sectional surface broaches for crank shaft bearings

Face A is machined by insert a, the two surfaces C by broach c, the flange diameter B by broach b, and the chamfer on B by broach f. The cylindrical portion of D is finished broached by insert a and the adjoining fillets by broach insert g. The faces at each end of the main bearing E are roughed and finished by broaches e and h, which also rough cut the bearing adjacent to the faces. The centre of the main bearing is roughed by broach j, and the bearing is finished over its entire width by broach k. The composite broach is 1,220 mm long, and in actual practice the broach units e, h, j and k will broach the other main bearings simultaneously with the bearings shown.

BROACH DESIGN

A broach is one of the most expensive single tools used for metal cutting. This is because its size requires expense in material, makes manufacture, including hardening, difficult and necessitates the cost of resharpening. Fortunately, the hydraulic broaching machine possesses features that help to prolong the life of a broach, i.e. by means of a pressure gauge on the machine the action or condition of a tool can be checked, indicating when resharpening is required. Also, when delicate broaches are being used, the relief valve can be set for a pulling pressure that will not damage the tool.

For some operations the length of a broach can be reduced by push broaching using a short sturdy tool, but usually the length of a broach comprises the length of the work plus the thickness of the faceplate plus the length of the adapter at the pulling end. In addition to the central cutting teeth, must be added say, four teeth for leading in to reduce shock, and say, 6 teeth for finishing, about 16 in all. On long broaches economies can be effected by building up the broaches by a series of high-speed steel rings, the remainder of the tool being of a cheaper grade of steel. This has the additional advantageous feature that, when some of the roughing teeth become undersized or damaged, the other rings can be moved along and new rings can be introduced. With all types of broaches when cutting steel, lubricants under high pressure to wash away chips are essential.

The total load on any broach section equals the total load divided by the area of section and, while an actual value for hardened high-speed steel is difficult to assess accurately, a value of 10,000 kgf/cm^2 can be used for tensile strength, but shock must be guarded against.

RESEARCH RESULTS

A survey from the Denver Research Institute with the results analysed by K.E. Miles indicates the following features.

The cutting speed was found to have little effect on the axial load, or on the size and shape of the hole produced. At high cutting speeds of 36 m/min or more, there is a tendency for chatter to be introduced.

Good surface finishes are associated with small values of cut per tooth, and a value of 13 μm was identified as the optimum. The number of finishing teeth should be at least equal to the maximum number of teeth in engagement during the cutting cycle. If extra broach life is required, this number should be increased by 40%.

The axial force in surface broaching increases as the cut per tooth increases, as does the normal force per tooth, but at a reduced rate. Axial and normal forces per tooth were mainly in proportion to the width of the surface being broached.

When broaching mild steel, flank forces on the tooth were influenced very little by the rake angle and were reduced when the minimum number of teeth were in engagement. If the cut per tooth is very small, the axial force was observed to increase while the normal force decreased and vice versa.

It was found that the least force per tooth existed if the summation of the pitches when the minimum number of teeth were cutting was just less than the length of the cut. For instance, if the broach has a pitch of 8.9 mm and the length of the cut is 19 mm, then 8.9 × 2 = 17.8 mm, thus slightly less.

For mild steel, the most effective rake was found to be 15 to 20°, for aluminium the value was 20° and for brass 5°. When surface broaching aluminium with a large rake angle, gullet space must be increased by 50% compared with cutting mild steel. When broaching mild steel, brass and aluminium, the changes in cutting speed up to 36 m/min did not change the type of chips produced. The length of cut, however, changed the shape of the coil.

Figure 3.13(a) shows a chart for surface broaching mild steel, 25 mm wide, the tool being high-speed steel and lubricant Mobil Vacmul 6R; Figure 3.13(b) shows the broaching of brass. The metal removal had a

Figure 3.13

A chart for calculating the cutting forces during broaching operations

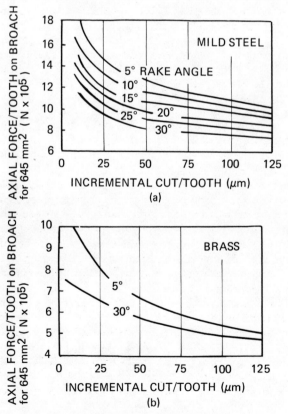

cross sectional area of 645 mm^2. Axial force is related to cut per tooth in each case, and it should be noted that the least force per tooth to cut a specified area occurs when the largest cut per tooth is employed. An increase in the rake angle also reduces the force per unit area, but for cuts in excess of 125 μm per tooth the force per unit area remains relatively constant. Cutting speed has no effect on the axial force required. From the experimental data collected, the following cutting force equation was obtained.

$$R = 1.45n \times \text{UTS} \times 0.94^{n/2}(1-\alpha/68)$$

where R is the resultant force on the broach tooth in newtons; n is the cut per tooth per 13 μm (t_1 μm) for a width W of cut equal to 25 mm, UTS is the ultimate tensile strength of mild steel in newtons per square millimetre$_1$, α is the rake angle and $t_1 \times W$ is the cross sectional area of cut.

Also, the resultant force equation can be used and here is based on broaching aluminium.

$$R = 1.55n \times \text{UTS} \times 0.95^{n/2}(1-\alpha\,80)$$

where UTS is the ultimate stress of aluminium in newtons per square millimetre.

4

ADVANCED
TRACER-CONTROLLED
COPYING SYSTEMS

EXAMPLE OF PRODUCTION BY COPYING

Prominence has been given to hydraulic systems in *Manufacturing Technology*, Volume 1, *Basic Machines and Processes*, and the subject is now extended to the copying of spherical and other more difficult profiles, with a discussion on the Heid electro-magnetic copying attachment. Two applications are shown; the first application is shown in Figure 4.1, indicating a large Morando lathe tracer turning a heavy connecting rod. A beam above the lathe carries a flat metal template, the stylus contacting this to produce a difficult curve on the workpiece. Copy turning makes good design feasible, examples being shown later, but the ability to

Figure 4.1
A tracer copying system for turning connecting rods
(by courtesy of Flli Morando & Co, Turin)

Figure 4.2
A set-up for copying impeller blades
(by courtesy of British Oerlikon Ltd, Birmingham)

produce such irregular curves enables danger points with alternating loads to be surmounted at relatively cheap cost and with a saving in weight of metal.

Another Heid tracer is shown in Figure 4.2, this time on a vertical milling machine, the workpiece being an impeller. This is mounted on a circular table with a dividing mechanism on the front of the rotating table to index for each blade section in turn. The tracing attachment is located on the left of the milling saddle; the stylus contacts a curved master of the profile required on the blades, the master being held in an ordinary machine vice on the milling machine table. (See Chapter 10 for the milling of an impeller using computer numerical control.)

THE HEID TRACER

A detail of the attachment is shown in Figure 4.3. It consists of a group of levers, the movement of these being transmitted to contacts which open or close electro-magnetic clutches. These drive the required feed motions in a direction selected by the contact finger A touching the template, and registering the profile variations by moving lever B. This lever is mounted on the equalizer C and acts on lever D by means of unit E. The lever D pivots vertically on stud F and determines the opening and closing of the contacts 1, 2, 3 and 4.

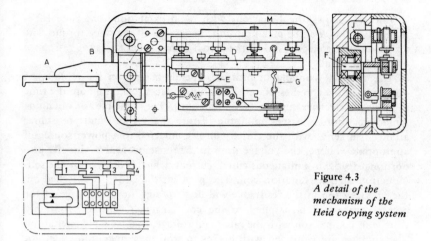

Figure 4.3
A detail of the
mechanism of the
Heid copying system

The spring G is connected to the contact holder M which turns horizontally on the fulcrum J towards the lever D. In this way at least one of the contacts can remain closed to operate a magnetic clutch and a selected feed motion. When the contact finger is in a dwell position, the distance between the contacts must be at least 0.05 mm and, when contacts 2 and 4 are in simultaneous operation, the distance between 1 and 4 must be 0.4 mm.

RAPID EXCITATION
The curves in Figure 4.4 show the torque increase under current, this increase being proportional to the current rise. The 'kink' in the curves is due to a sudden change in self-induction caused by the axial movement of the armature and it represents the moment of clutch engagement. Curve A shows the behaviour of the clutch without forced excitation in which engagement does not take place until 0.032 s, the torque transmitted at the moment of engagement being only 18% of rated clutch torque.

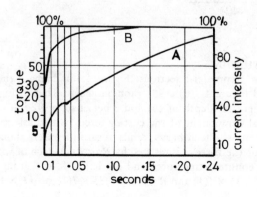

Figure 4.4
Curves showing
torque features
of Heid clutches

Curve B shows the behaviour with rapid excitation. The time lag is reduced to 0.009 s, the pick-up torque being 68% of the rated torque. The resistance of the electric link in the clutch circuit is ten times the resistance of the clutch coil.

The accurate copying of profiles depends mainly on the rapidity of operation of the clutches. Rapid excitation reduces to a minimum the time loss due to electrical inertia (self-induction). The principle for obtaining rapid excitation is based upon the feature that, if an ohmic resistance arranged in series with the clutch coil is connected to a power source of appropriate voltage, the voltage drop in the resistance will be directly proportional to the instantaneous clutch current, which is delayed by the self-induction in the excitation coil. The moment the clutch is switched on, the voltage drop in the resistance will be zero and the full voltage of the power source will be applied to the coil terminals. The higher supply voltage in comparison with the rated coil voltage, enables rapid excitation. The clutches are equipped with brakes to stop their operation when not driving the mechanism.

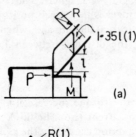

(a)

Figure 4.5
Examples of cutting tools used for copy turning

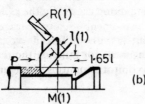

(b)

TOOLS FOR COPY TURNING

Figure 4.5 shows two commercial sections, Figure 4.5(a) being a rhomboidic $58°$ insert and Figure 4.5(b) a $55°$ rhombic insert. That in Figure 4.5(a) permits a 40% greater depth of cut and is better supported. It indicates how a horizontal component of the cutting force, particularly if the depth of cut is small and the feed heavy, tends to turn the insert about M or $M(1)$. The action of the force P is resisted by R or $R(1)$, each of which is the resultant of a counteracting force and a friction force. From the diagrams, $P \times l = R \times 1.35 \times l(1)$ and $R = P \times l/1.35 \times l(1)$; also $P \times 1.65$

$\times I = R(1) \times I(1)$ and $R(1) = P \times 1.65 \times I(1)$. Therefore $R(1)R = 1.65 \times 1.35 = 2.23$, i.e. the counteracting force for the rhombic insert is 2.23 times as great as the rhomboidic one. Nevertheless, the rhombic shape is more suitable for turning a component with the taper section shown.

COPYING OF SPHERICAL SURFACES

Spherical surfaces, either external or internal, can be produced on various components, as, for example, ball and socket joints, lapping plates, and other convex or concave contours. Figure 4.6 shows a graph of the copy

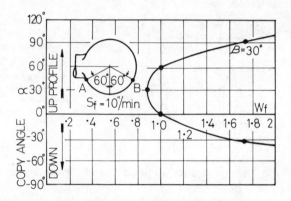

Figure 4.6
A graph showing the copy angle for spherical surfaces

angle α against the workpiece feed W_f when the saddle feed S_f is unity. From the graph, W_f can be found at positions A and B when the saddle feed is, say, 10 in./min. Thus, at position A, $W_f = \alpha$ and, at B, $W_f = 10$ in./min.

Let us consider some examples: in Figure 4.7(a) the cylindrical template has four centre holes, two being used for grinding the template and the other two off-set from the axis, for mounting the template between

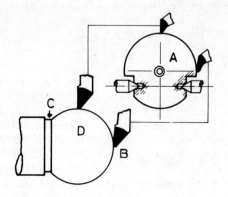

Figure 4.7
An example of a copying sphere from a cylindrical template

centres as shown. Copying starts at point B and terminates at C on the workpiece D. The front tool post is used in the first operation for turning the shank and cutting the groove at C.

Figure 4.8 shows the setting for turning a sphere with a radius greater than the stroke of the copying slide. Movement of the stylus along the arc A reproduces a sphere which is machined in two settings. In the first operation, left hand, the hydraulic cylinder of the tool slide is in the fully forward position, and the stylus contacts the template at point B and traces an arc to the outside diameter of the component. In the second

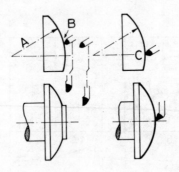

Figure 4.8
A setting for a turning sphere of large radius

operation, right hand, the stylus is set on point C, and the turning tool is set on the work centre.

By this method it is possible to machine large spherical or other curved surfaces by hydraulic copying, the size of the step depending on the radius required on the workpiece and on the stroke of the copying slide.

Internal spheres can be produced as in Figure 4.9 using cylindrical templates; the template A is made from a flat ring, the internal diameter corresponding to the diameter of the component required. Centre holes for mounting are provided, together with a horizontal flat B to act as an additional guide surface for the stylus so that a component is machined without traversing the hydraulic cylinder to its full forward position.

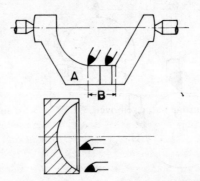

Figure 4.9
The copying of an internal spherical surface

Figure 4.10
*A copying method
using two tools in
one rest. The spindle
reverses when tool
B has completed
cutting, and before
tool C commences*

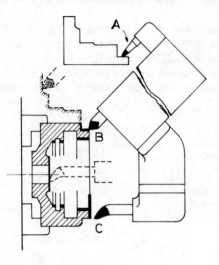

TWO TOOLS IN ONE REST

A system of using two copying tools, as in Figure 4.10, where both are mounted in one rest, can result in a reduction of machining time, because it combines two operations. Controlled by a single tracer, two tools are brought into operation one after the other. The stylus A follows the outline of the template and controls the movement of the tool rest. While the first tool is cutting, the tool C also travels but without touching the work until it is brought into engagement with the work, when, on completion of cutting, the first tool leaves the workpieces, following the same path as the second tool.

To reduce the length of the cutting traverse, and hence the production time, multiple tools can often be fitted in the copying rest, as shown in Figure 4.11, to limit the traverse to about the length of two shoulders

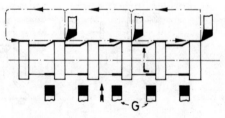

Figure 4.11
*Copying tools used
in conjunction with
tools in a front rest*

instead of six. The shape produced on the component is that shown along the line L, but, by mounting a series of grooving tools G in the front rest, the surplus metal required to be removed to give square shoulders can be accomplished simultaneously by a short inward traverse of the tools.

Similarly, where considerable metal removal is required, time can often be saved if several tools in the front rest are suitably spaced with regard to distance apart and depth so that, by a short longitudinal traverse of the rest, the main part of the component is produced roughly to shape. The copy tool is then used to give the final shape, while the life of this tool is increased by the reduction in metal removal.

Figure 4.12 shows another example where a Kosta driver is used with multiple tools in the front rest. Thus complete machining of a shaft can

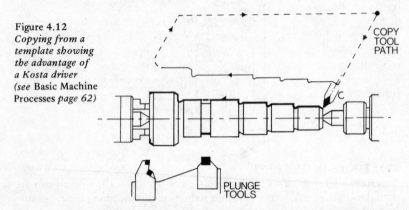

Figure 4.12
*Copying from a
template showing
the advantage of
a Kosta driver
(see* Basic Machine
Processes *page 62)*

take place without a second setting. The copying tool C can traverse the full length of the component under template control, and then the tools in the front rest complete the machining by undercutting, bevelling and finishing the large diameter to the required thickness.

USE OF DUPLEX TEMPLATE
An example (Figure 4.13) of the machining of handwheels shows the use

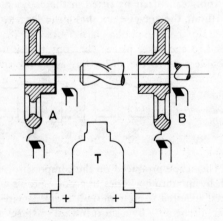

Figure 4.13
*The use of a duplex template for
machining handwheels*

of a double-sided template, the first operation A using the left-hand side for the traverse of the stylus and tool control, while for the second operation the template T is turned over to control the tool traverse for completing the outside diameter and facing the boss. The workpiece is now held in the lathe chuck on the previously machined boss. An attachment fitted to the front tool post is available for holding boring tools for machining the bore.

Figure 4.14 shows that, where opposite shoulders are normally bored in two settings, it is sometimes possible to use one setting only if a duplex tool holder and two templates are used. Tool T_1 feeds from left to right as controlled by template 2, while template 1 controls the movement of tool T_2 traversing from right to left. This method of copying can be used for either internal or external machining of many components.

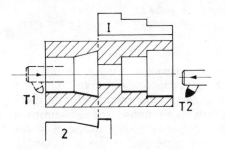

Figure 4.14
An example of copy boring using two tools

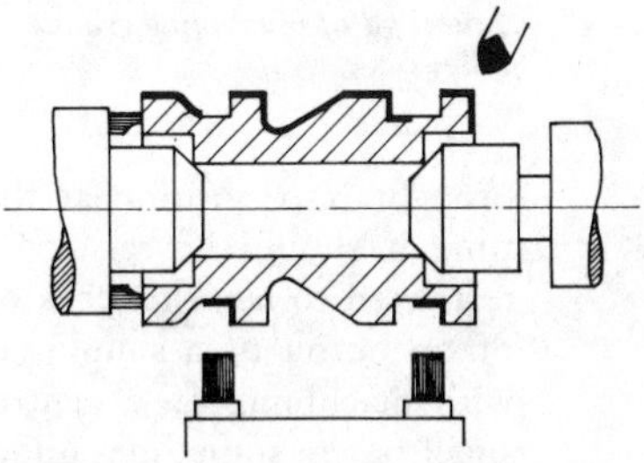

Figure 4.15
A set-up for a copying thimble using two rests

A set-up for the copying of a clutch thimble is given in Figure 4.15. The diagram shows how the component is mounted so that no restraint in copying from end to end takes place. The copy tool is unable to produce right-angled shoulders down the slope, but again the advantage of being able to use the front tool post is shown in that the machining can be completed by use of the two grooving tools.

Figure 4.16 gives an example of the machining of grooved pulleys for rope drives. By the use of several templates it is possible to machine the outside diameter of the boss and some of the other internal contours if balancing is required; however, if we restrict the example to the machining of the vee grooves, a template can be used to face the end and to produce the three steps by a single-point tool as at B. The advantage of this is that it reduces the amount of metal to be removed in producing the vees using the front rest with the form tool E. It can be seen from the examples given that, if sufficient quantities of work warrant it, almost all sections of a given component can be copy turned by using several templates in succession. In some cases special holders can be provided so that these can be

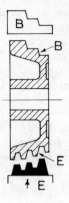

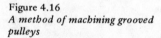

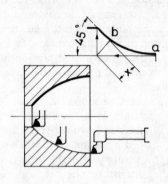

Figure 4.16
*A method of machining grooved
pulleys*

Figure 4.17
A procedure for copying a deep profile

brought into position at the required sequence of cutting. At the same time it should be realized that the advantages of copy turning are not restricted to large batches of work or complicated contours, for the speed of operation of a simple copying lathe is so pronounced in comparison with machining on a centre lathe with normal tooling that, even with a small batch, sometimes under five components, great savings in production costs can be expected. (See Chapter 11, Figure 41.)

MACHINING A DEEP PROFILE
There are limitations to the depth of profiles that can be machined, generally depending upon the size of the lathe and on the length of travel of the copying slide. In some instances this can be overcome by special adaptations or by the rearrangement shown in Figure 4.17. The copy boring tool commences traversing under template control and, if the diameter variation is greater than the full stroke of the copying slide times the sine of the angle at which the slide is set, sliding feed is given to the saddle for moving the stylus between points a and b. This motion is then stopped and the surfacing feed is used to complete the operation.

During movement of the saddle, the copying slide is traversed in a direction away from the work under template control, up to the maximum travel equal to x, for machining part of the profile. When the surfacing feed is engaged, the slide is moved in the other direction to complete the operation.

To revert to the machining of spheres, the production of a ball and socket joint was demonstrated to me on a Harrison copying lathe. The ball 4 in. in diameter comprised a $120°$ section of a sphere, and this and the socket were copied from sheet metal templates ground to the section required. With a light coating of oil, although the ball would swivel in the socket under the lightest finger pressure, the adhesion was such that the units could not be pulled apart by ordinary manual force; when using both hands to pull on the ball, the result was simply to lift the heavy socket with it, thus showing the accuracy of fitting all around the spherical section. A further proof was demonstrated by marking with Prussian blue.

QUICK-CHANGE TOOL POSTS

With quick-change tool posts, tools can be repeatedly mounted and removed with the tool position precisely maintained. There are tool holders available for every operation with reproducibility better than 0.0003 in. and accuracy is not affected by wear. Setting time is reduced because rapid adjustment for height is obtained without packing and tools are reground in their holders. Where long runs are involved, spare tool holders can be made available so that delays for tool sharpening are avoided.

DESIGN FREEDOM

Copy turning makes weight saving and good design possible without additional costs. Walls of uniform thickness with corresponding internal or external forms can be produced without difficulty. On parts subject to heavy alternating loads, the profile of the part is of the utmost importance, and Figure 4.18(a) shows how the danger points indicated on the section below the centre line can be avoided without the use of expensive form tools.

To design a part as in Figure 4.18(b), with the ideal profile based on the material being used, may cause machining difficulties. As a result the section below the centre line is often used, but with copy turning the ideal profile to resist stress concentration, shown above the centre line, can be produced without any difficulty.

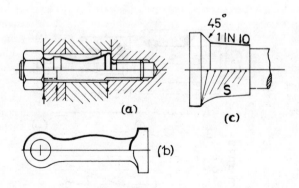

Figure 4.18
An example showing how copying assists design freedom

The double-cone-type bearing is used on a range of machine tools varying from boring machines to watchmakers' lathes. This bearing is an approximation to the Scheile curve, in which it has been shown that an end thrust bearing made to this curve will have uniform wear at all points. Unfortunately, to machine this curve to suit both spindle and bearing has hitherto been very difficult and the approximation shown above the centre line of Figure 4.18(c) has been accepted. With a copy lathe, however, the machining of the two mating components is simplified, and the genuine curve S as produced by Scheile in his experiments is easily copied.

COPY TURNING VERSUS MULTIPLE-TOOL TURNING

The main purpose of copy turning is to eliminate the complications of expensive multi-tool set-ups and, although some examples have been given to show that multi-tooling can assist copy turning, these have been limited to a few tools to complete an operation previously copy turned and there is no doubt but that development will proceed along these lines. This is feasible, and indeed useful, in making use of both tool posts provided on the copy lathe.

A comparison of the accuracy obtained by single-tool copying and a multi-tool set-up shows that greater accuracy can be obtained with the first system. Tests on the same component, using identical speeds and feeds, and cutting tool or tools, have been carried out to measure the stress during machining operations. These were measured by a low-inertia system comprising a capacitance pick-up and an amplifier, with the results recorded on an oscillograph.

It will be appreciated that stiffness or rigidity of a shaft is the main feature in ensuring accuracy of machining, but the number of tools in operation and the difficulty of preventing vibration in a tool holder of some length is a factor. Figure 4.19(a) shows a typical set-up for multiple

Figure 4.19
A comparison of the accuracy for single and multi-tool turning

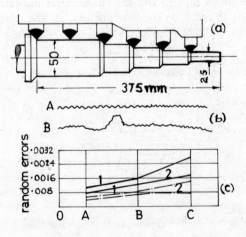

tooling, whereas Figure 4.19(b) shows the results on an oscillograph A obtained from copy turning and B from turning with multiple tools.

It can be seen that the latter is accompanied by considerable variations in the elastic displacements along the length of the shaft; thus the dimensional error is much greater than with single-tool copy turning. The random errors in rough turning shafts of the dimensions given reached 0.0028 in. compared with 0.0016 in. in copy turning; in finish turning the errors were 0.0016 and 0.0008 respectively.

With a multi-tool set-up as shown, dimensional errors can develop in inaccurate tool setting or uneven wear amongst the various tools. Consideration must also be given to the power requirements for heavy metal removal. Records taken on a moving coil wattmeter frequently indicate that a driving motor is being overloaded with certain set-ups on a multi-tool lathe; this necessitates a less satisfactory regrouping of certain tools to balance the load more nearly. A further factor to be taken into consideration with multiple tooling is the work deflection caused by the cutting pressure, and this may necessitate the fitting of steadies and may thus increase the setting-up time.

A simple copy lathe may be preferable on first cost alone and may well score on the time taken for machining, because the initial tool set-up and subsequent grinding and resetting of several tools on a multi-tool lathe is expensive when compared with single-tool maintenance.

In multi-turning of shafts similar to the one shown, when rough turning at a depth of 2 mm, the random errors in dimensions were 60% of the tolerances; this was reduced to 30 to 40% for single-tool copying. For finishing cuts these dimensions were reduced to 25 to 40% for multi-tool and to 20 to 25% of the tolerances for copy turning.

Figure 4.19(c) shows a typical graph, indicating the instantaneous dimensional random errors in relation to shaft stiffness in multi-tool turning (curve 1) and single-point copy turning (curve 2); the full curves indicate rough turning and the chain curves indicate finish turning. The letters A, B and C on the base line indicate the relationship of length to diameter: A is less than 10, B is equal to 10 and C is greater than 10.

EXAMPLES OF INTRICATE COPY TURNING

Some examples of components produced on Harrison copy turning lathes are now described.

Six-spline broach

A six-spine broach (Figure 4.20) is of high-speed steel; the finished size of the teeth are ½ in. (12.7 mm) in diameter. Thus the small diameter relative to the length requires considerable care in machining. The first operation is to turn a portion $7/16$ in. in diameter and $2\,3/8$ in. long from from the original which is $5/8$ in. in diameter; this operation takes 1¼ min. In the second operation, the bar is held in a three-jaw chuck on the machined $7/16$ in. diameter; the rest of this diameter, as well as the broach

Figure 4.20
An example of the copy turning of a spline broach

section and end, is turned over its entire length. For this purpose a travelling steady is fixed in front of the saddle and the cutting tool is extended to a position in front of the steady. Copy turning of the broach section is completed in one traverse at 818 rev/min with a feed of 0.006 in./rev (0.15 mm/rev). The floor-to-floor time for this operation is 6 min.

Alumina ceramic cone

Copy turning is proving its value in machining some of the newer materials, and the example of an alumina ceramic cone (Figure 4.21) shows how an intricate section with a wall thickness of only $\frac{1}{32}$ in. can be produced by

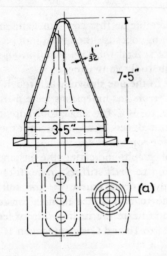

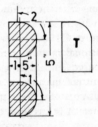

Figure 4.21
The copy turning and boring of an alumina ceramic cone

Figure 4.22
The procedure in copying a plastic waste pipe

both internal and external copy turning. Material removal of 0.25 in. on each surface is required and, although the external contours are not difficult to produce, the internal boring operations to leave an even wall thickness on the taper requires a machine entirely free from vibration and a copying system with a high degree of accuracy.

The chain lines show the rigid boring bar employed for the internal machining, the end of the bar being cut away so that it is possible to produce a rounded end in a very limited space. The cutting speed employed is 2,000 rev/min using a feed rate of 0.003 in./rev (0.07 mm/rev), giving a floor-to-floor time of 6 m.

Die for plastic waste pipes

The material for the die for plastic waste pipes (Figure 4.22) is mild steel and machining is carried out at 289 rev/min with a feed rate of 0.005 in./rev (0.12 mm/rev). A template shown at T is used. In the first operation, face 1 must be profile turned using an internal copy tool bar. For this operation the tool must be set so that the cutting edge traverses parallel before producing the radius, thus ensuring that the limit of the radius has been passed. In the second operation, using a standard profiling tool, copying of section 2 takes place firstly by turning a portion of the parallel section and then by profiling the curve so that a correct blending of the two radii takes place. The floor to floor time for the operations is 12 min.

COPY MILLING AND DIE SINKING

The limitation to copying by mechanical means is that, to maintain contact between a roller and a master a heavy load must be imposed between the two components, either by weight and chain or by spring pressure. Not only does this feature produce high friction between the contacting parts resulting in wear, but the main feature is that there is a definite limiting angle of about 30° beyond which it is almost impossible to produce an accurate copy owing to the heavy increased load which takes place as the angle of climb increases. No such restriction takes place on a machine fitted with tracer-controlled copying equipment, for the contact pressure between stylus and copy may be as small as a few ounces-force.

Consider an example of machining the contours of rotor blades for air blowers; with mechanical methods four operations are required to cover the contour because it is impossible to have a rise in the former plate which is so steep that a tool can reach to the centre line. With tracer control, however, the component can be machined in two operations.

An example is shown in Figure 4.23 which indicates the copy milling of special heavy foundation bolts on a horizontal machine. The hydraulic copy unit is seen on the left, the stylus carrying a roller contacting the master on the machine table. The workpiece has a steep rise section which is produced by the roller actuating the tracer valve and thereby the rise and fall of the machine table. The bolt tapers along its length, but this is taken care of by the shape of the milling cutter.

360° TRACER MECHANISM

The simpler 180° tracer milling equipment has been described in *Manufacturing Technology, Basic Machines and Processes*, page 151, but for milling many dies and moulds the tracer through the master must control the feed rate of two slides moving at right angles to each other. This type

Figure 4.23
Form milling by hydraulic tracer control

of tracing permits the reproduction of shapes in a horizontal plane through 360°.

Figure 4.24 shows an example of copy milling the slots in special cylinders, the workpiece being shown on the left under the milling cutter with the master on the right under the stylus. Both workpiece and master are held in a fixture which is rotated to bring the slots, in turn, under the cutter and stylus, the latter controlling the table movements.

The 360° mechanism is shown in Figure 4.25. The tracer valve corrects the direction of movement only, the proportion between the two feed components required to obtain this direction as a resultant being governed by separate feed control valves. The tracer stylus is eccentric by 0.015 in. to the tracer axis. Geared to the tracer body is an eccentric feed cam which controls the settings of two feed control valves at right angles to each other, one controlling the transverse slide cylinder and the other controlling the table traverse cylinder.

If the direction of the tracer eccentric is such that it interferes with the master, the tracer overdeflects, lifting the piston valve through the ball and setting in motion a hydraulic motor geared to both the tracer body and the feed cam, causing the tracer to rotate anti-clockwise. This reduces the tracer deflection and brings the feed cam to a position to adjust the feed components to give the correct directions. In this way, the motion of the tracer tends always to be in a direction tangential to the curve of the master. The rate of feed is controlled by the amount of eccentricity of the feed cam. In this type of mechanism, the servo can be extremely sensitive, since any 'dither' in the servo mechanism will have only a secondary effect on the slides. It has been found that a vibration of low amplitude and high frequency will not affect the slides and at the same time will increase the accuracy of tracing.

Figure 4.24
A 360° tracer arrangement for milling ports

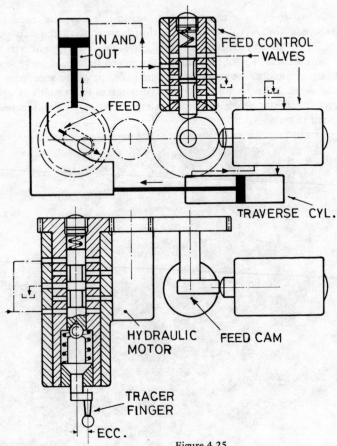

Figure 4.25
A detail of tracer valves for copy milling

PROFILING BY OPTICAL TRACING FROM DRAWINGS

Contour milling direct from template lay-outs has been used in which an optical system scans the lay-out and projects an image of the line, enlarged seven times, onto a glass screen. By manipulation of hydraulic controls, the operator keeps the image of the follower in contact with that of the line and, as the lines are followed, they are reproduced by a milling cutter on the surface of the workpiece. Although the large magnification reduces any following error, a more accurate system is a system which used photo cells.

DRAWING REQUIREMENTS

The Linemaster system (Andrew Engineering Co., Minneapolis, U.S.A.) uses polyester resin sheet material on which line drawings are made by a

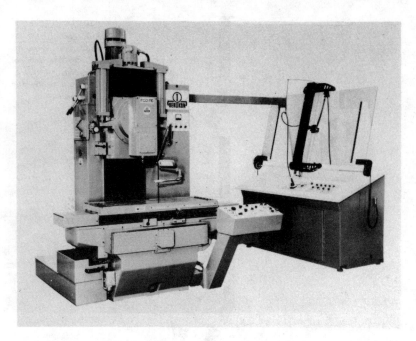

Figure 4.26
A milling machine controlled by the photo-cell action from a drawing

ball point pen with black ink, using a parallel ruling straight edge and set squares. A 1 mm ball pen is suitable, and pens should be adaptable for use in compasses.

The drawing is taped to a back-lighted table (Figure 4.26), and the tracer head which controls a milling or die-sinking machine works by photo-cell action. The head is moved forwards towards the outline to be traced and, when it reaches the line, it locks on and follows the drawn path. The tracer system picks up signals from the sensing head and, through servos, guides the milling machine along the same path.

The sensing head stays 'on course' by balancing its intake of light. Two photo-cells in the head straddle the line on the drawing and, if the line changes direction, more light enters one cell than the other. This imbalance sets up an error signal in the tracer system, which rotates the head until balance is restored. As the sensing head moves in the new direction, so does the cutter on the milling machine.

For hole locations travel is along an intersecting line as before and, when it reaches a cross line indicating a hole centre, a different photo-cell from the line-following photo-cells senses the cross line and stops the head and machine while the hole is drilled.

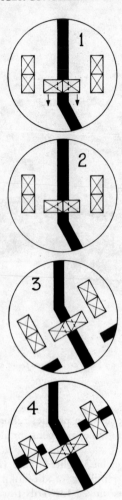

Figure 4.27
A diagram of the photo-cell action of a Linemaster system (by courtesy of Andrew Engineering Co, Minneapolis, USA)

THE PHOTO-CELL ACTION

Figure 4.27 shows in diagram form how the system works. In diagram 1 the photo-cells are indicated by the diagonal rectangles, while the heavy line represents the line being followed on the drawing. In this case the cells balance on the line in a forward direction. In diagram 2 there is light imbalance to the cells, the change in line direction causing rotation of the tracer head. In diagram 3 the cells restore the balance so that the head moves in the new direction, and cross lines begin to appear; in diagram 4 the cells balance on the cross line and stop the tracing action.

Figure 4.28 shows a range of examples machined by optical tracing, indicating that the system provides a means of producing intricate work

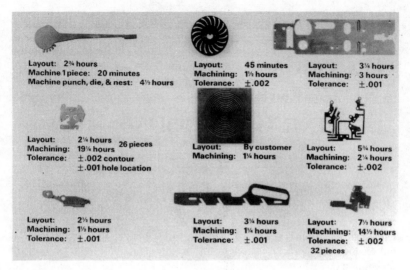

Figure 4.28
Examples of workpieces produced from line drawings

quickly and economically. Preparations for the process are simple and, because of its graphical nature, optical tracing compares favourable with NC for prototype work. Making a drawing is simpler and faster than programming a part and punching a tape, even if a computer and related equipment are available. The drawings may be of various scales up to 20:1 with electronic division between the reader head and the machine drives.

Further examples of various copying systems are given in *Manufacturing Technology*, Volume 1, *Basic Machines and Processes*, Chapter 7.

5

ELECTRICAL AND CHEMICAL MACHINING METHODS

A number of electro-erosion processes are available for removing metal to produce simple or intricate contours in metals previously difficult to machine by conventional methods. Two prominent means are electro-chemical machining (ECM) and electrical discharge machining (EDM). Other means include electron beam machining (EBM) and laser beam machining (LBM).

ELECTRO-CHEMICAL MACHINING
A main advantage ECM is the high metal removal capacity. Most steels can be machined at the rate of 990 cm³/h on a 10 000 A capacity machine. Also, most conductive materials, regardless of their hardness, strength and machineability, can be processed. It is essentially a stress-free process since the electrolyte pressure is the only force involved, so fragile parts and stress-sensitive alloys can be machined. Temperatures are limited to about 120°C, the boiling point of the electrolyte, so that alloys with low melting points cannot be machined.

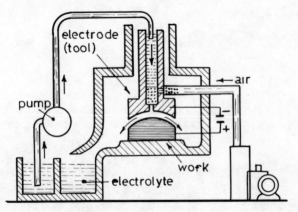

Figure 5.1
A diagram showing the principle of ECM

PRINCIPLE

ECM is based upon the electrolytic dissolution of the workpiece, which serves as a positive electrode. As shown in Figure 5.1, when the electrolyte is impelled past the negative or tool electrode at high speed and a large current passes, the positive work piece undergoes dissolution in accordance with the shape of the electrode. There is no tool electrode wear, so the same electrode has a long life, and neither speed nor working accuracy is affected by the hardness of the material. The operation produces little heat and has no effect on the work which can be heat treated before machining. Any electrically conductive material can be selected for the electrode, but the most suitable materials are brass, steel, copper, stainless steel and, in special cases, carbon.

ELECTROLYTE

The electrolyte provides a current path, flushes away metallic hydroxides and removes heat. The most common electrolyte is brine (25% salt) because of its low cost and low toxicity, but a sodium nitrate solution gives a machining rate 0.6 times higher than that of brine. The sodium nitrate is mixed in the ratio of about 1 kg per 10 litres of water. As shown in the diagram, if compressed air is introduced into the electrolyte to agitate it or to eliminate stratification, machining accuracy is improved, especially at the bottoms of workpieces.

The effect of tool feed rate and supply voltage on accuracy is that the predominant resistance to current flow in ECM is that due to the electrolyte, and hence the current rate may be expressed as

$$i = \frac{V_s}{R_e} = \frac{A V_s}{r_e}$$

where V_s is the supply voltage, R_e is the resistance due to the electrolyte, r_e is the specific resistance of the electrolyte, and A is the area exposed to electrolysis.

An expression for the gap width δ between the tool and work surfaces can be obtained from the above formula; the rate f_t for the tool feed is given by

$$f_t = \frac{V}{A}m = \eta \frac{ei}{pA}$$

where A is the area of work exposed to electrolysis and $\frac{i}{A}$ is the current density. The equation

$$\delta = \frac{V_s}{r_e p f_t}$$

shows that an increase in f_t causes a decrease in δ, whereas a small gap means a higher accuracy of reproduction and therefore a higher metal removal rate with greater accuracy. Also, an increase in V_s increases δ and reduces accuracy. The supply voltage generally used ranges from 5 to 20 V d.c., the lower value being for finish machining and the higher value for roughing.

Applications for ECM include trepanning operations, blind holes, cavities in forging dies, turbine blades, internal splines and surfaces of revolution. An example of trepanning in 80 tonnes steel shows that 1800 mm^3 of metal was removed in 3 h.

CHEMICAL MACHINING

In chemical machining, metal is removed from a workpiece by immersing it in a chemical solution that is capable of metal dissolution. The component is coated with a masking compound and the areas to be 'machined' are exposed by cutting and peeling away the mask.

The depth of 'machining' is controlled by successive immersion of the work and by removing progressively further areas of making compound. A typical process cycle comprises degreasing of the work, application of masking compound, cutting this away where required and immersing in the chemical fluid.

Vinyl, neoprene and butyl-based masking materials are the most widely used and are applied by spraying or dipping, to provide coatings up to 0.4 mm thick.

This chemical milling generates heat, and, since the rate of dissolution is proportional to the temperature of the etchant at the 'attacked' surface, effective agitation and circulation of the solution is essential.

The workpiece shown in Figure 5.2 is typical and is an end pre-load for

Figure 5.2
*A typical example
of a workpiece
produced*

a precision bearing assembly made from spring-hard stainless steel. In general, for aluminium alloys such as Alclad and Hiduminium the etchant used is sodium hydroxide at $90°C$. For stainless steel, the etchant is an acid fluoride at $65°C$ or hydrochloric acid. For titanium, acid fluoride at room temperature is used and, for Nimonic alloys, hydrochloric or nitric acid at $60°C$ is employed.

ELECTROLYTIC GRINDING

A metal or graphite wheel is used for electrolytic grinding; this has abrasive

Figure 5.3
*A diagram showing
the electrolytic
grinding system*

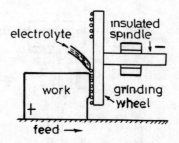

particles, and the wheel and its spindle are insulated from the rest of what is almost a conventional grinding machine; current is passed from the wheel to the workpiece (Figure 5.3). Actual contact of the wheel and work is prevented by the insulating abrasive particles which protrude from the surface of the wheel and leave a 0.02 mm gap into which a stream of electrolyte is directed. The work is pressed against the wheel and some material is removed abrasively, but 80 to 90% of the total removed is by electrolytic action.

Electrolytic grinding permits higher metal removal rates, particularly on hard materials, and also economies in the use of expensive abrasive. It is widely used with metal-bonded diamond wheels in the grinding of cemented-carbide tools, while carbon-bonded solid abrasive wheels are used for stainless and hardened steels. Electronic honing is another extension for faster metal removal, the advantages being similar to these claimed for grinding.

SPARK-EROSION MACHINING

The spark-erosion process requires two conductors, the electrode and the work, separated by a spark gap and both immersed in a dielectric fluid. If an increasing voltage is applied to the electrodes, breakdown (ionization) of the dielectric occurs, resulting in a spark. It is this spark that causes metal removal.

The use of the dielectric fluid facilitates the control and continuation of the spark while concentrating the spark energy on to a small area of the work for maximum erosion effect. Secondary functions of the dielectric are to wash away the swarf and to keep the workpiece cool.

The electric discharges occur usually between high spots on the electrodes where the field intensity is greatest. Erosion of each high spot (with erosion of a greater degree on the workpiece) creates further high spots which are in turn removed. This successive removal of high spots continues until a complete area has been eroded into the workpiece. To maintain continuous sparking, the tool electrode is fed downwards by an automatic servo system that uses the spark gap as a reference for the feed control. The required depth is obtained by a manually set cut-off switch.

DESIGN TECHNIQUES

Figure 5.4 shows the principle of EDM or spark machining, the workpiece

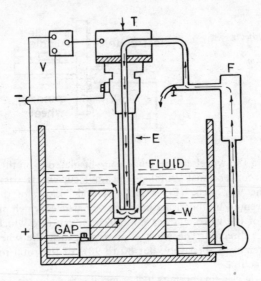

Figure 5.4
The principle of EDM

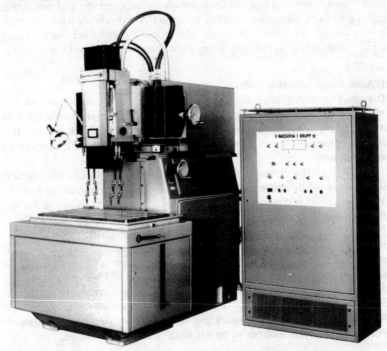

Figure 5.5
Spark machining on a Scheiss–Froriep UK machine

being indicated by W and the servo control by V, while T shows the tool feed. E is the negative electrode shown immersed in the dielectric fluid which may be paraffin, transformer oil or white spirit. The shaped tool is often of copper or tungsten-copper. The pump circulates the fluid through the filter F after drawing from the reservoir.

The feed rate is continuously controlled to maintain a fixed breakdown voltage across the gap, corresponding to a mean electrode-workpiece separation of 0.02 mm.

Figure 5.5 shows the design of a machine by Schiess-Nassovia (Schiess-Froriep (U.K.) Ltd). This is a two-channel machine, but up to 10 can be supplied if required. The erosion depth is pre-selected by setting slip gauges and, when the required depth is reached, a precision switch interrupts the supply to the spark gap. The quill is then retracted and clamped by a brake.

A diatomite filtration system is used, the filtration medium being flushed through a large number of tubes made from braided wire. The dielectric is thus cleaned while the diatomite layer flows from the outside to the inside. When the pressure difference has reached a certain maximum level with rising contamination, the filtration medium (now loaded with erosion particles) is flushed into a sludge tank. The smallest tank in the range of machines has a capacity of 980 litres, and the largest 15,000 litres.

PULSE GENERATORS

Because the voltage needed to sustain a discharge is lower than that needed to initiate it, it has been found advantageous to employ pulse generators which superimpose an ignition pulse onto the working pulse. This ensures regular spark discharge with consequent regular metal removal; without the ignition pulse, irregular sparking may occur at intermittent intervals which necessitate a wider spark gap to avoid contact between the work and the electrode. An optimum spark gap being employed permits optimum metal removal rates with better flushing of eroded particles and work surfaces finishes of 0.4 Ra μm.

Some examples of components produced by spark erosion include drilled holes in gas turbine blades in Nimonic 108 produced in 7 min, air transfer tubes, flame tubes with radial holes and 24 slots in Nimonic material, forging dies, die-casting dies, car body dies, stamping tools, moulds for glass manufacture and a wide range of other components in materials difficult to shape by other means.

ULTRASONIC MACHINING

Ultrasonic drilling is based upon a principle in which high-frequency electrical oscillations are produced in an electronic generator and are supplied to a piezo-electric or magneto-strictive transducer which converts them into mechanical vibrations. The amplitude of the resulting vibrations is increased by the use of a so-called velocity transformer, to the end of which is attached a cutting edge or bit.

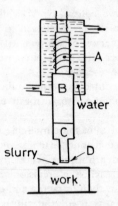

Figure 5.6
*The action of
ultrasonic drilling*

Figure 5.6 shows the basic construction of an ultrasonic drill, where A is the magnetostrictive transducer, B the half-wavelength coupling rod, C the velocity transformer to increase the amplitude of vibration and D the tool brazed to the velocity transformer. The tool is not directly applied to the material being cut but acts through the medium of an abrasive slurry; the actual cutting is carried out as a result of the fact that this abrasive is violently driven into the material by the reciprocating action of the bit. The drill is not limited to the production of round holes, and complicated shapes can be obtained by using a bit of the required form. Abrasives used vary according to the materials to be worked, and for most purposes silicon carbide, boron carbide and aluminium oxide of various grain sizes are suitable.

Machines for heavy operations operate at a frequency of 20 kc/s, and power up to 2 kW is provided by the electronic generator. Holes can be drilled with a diameter from about 1.5 mm to 60 mm, and the range of materials worked includes glass, ceramics, tungsten carbide, germanium and precious stones. The tool should be hard, but not brittle and in operation is vibrated against the work surface with an amplitude of about 0.05 mm at about 20,000 cuts/s. The loose abrasive is thus hammered into the work surface, material being removed by a chipping action, but on a microscopic scale. The frequency is above the upper limit of audibility, so that the process is silent, while thermal cracks associated with heat-treatment processes are non-existent because the small amount of heat generated is entirely local.

The maximum cutting speeds in materials such as glass and softer ceramics are about 2 cm/min. Speeds are a factor of 5 less in materials such as fused alumina and gem stones. Dimensional accuracy of the order of ± 0.005 mm with a surface finish of 10 μin. is usual, and the wear of the tool head is relatively slight, except when cutting die steels and sintered carbides; in these cases for high precision it may be necessary to take both roughing and finishing cuts.

THE LASER IN ENGINEERING MANUFACTURE

THE BASIC BEAM

The laser is an oscillator and must be supplied with power from which to select its own characteristic form of energy and to store it. The store is topped up from the power source, thus compensating for the small amounts of energy that are lost by allowing some energy to emerge usefully and by other less useful side effects. The discharge tube contains helium and neon, and it creates clear coherent light; it is this coherence of the beam from a laser that makes it significant to the engineer.

The laser is pre-eminent in the field of very accurate measurement, but developments have now increased its use towards cutting metals by combining a stream of oxidizing gas. With the introduction of the carbon dioxide laser of the molecular gas type, output powers of several hundred watts are obtained readily. The output radiation is in the infra red region of the spectrum and a wavelength of 10.6 μm is typical.

A practical advantage of the carbon dioxide laser is that machining operations can be viewed through windows of Perspex or other materials that are opaque to the wavelength, the 10.6 μm wavelength giving a good transmission through haze and smoke.

CUTTING, WELDING AND DRILLING

There are many forms of carbon dioxide lasers and, with slowly flowing gas in the discharge region, outputs are in the middle range of power from 5 to 1000 W. These lasers are used in cutting operations. Gas moves through the tube faster when the bore is increased and, as this reduces heating of the gas, so more laser power is developed. Units based on this concept are known as fast axial flow lasers, and their outputs are from 500 to 2000 W, making them suitable for deep penetration welding.

To reduce the impedance to the gas flow and to allow a large cross section to be used, the discharge can be struck at right angles to the gas flow. Output from this type has reached 25 kW, the power level making surface hardening and surface alloying, as well as deep penetration welding, possible. The laser beam has no inertia so that complex shapes can be cut at high speed. Also the beam has no mass, so instantaneous changes in direction can be made on command. An intense laser beam is capable of cutting difficult materials such as titanium or high-speed steel. Clamping of the workpiece to resist the cutting force is not necessary, and the laser beam does not suffer from wear as does a metal tool.

Highly accurate components can be made with simple guidance or NC. Cut edges are square to the surface with little slag, while the heat-affected zone is only about 0.07 mm wide. The success of the low-powered laser has been due to the advent of the gas nozzle; one design is shown in Figure 5.7. The advantage is that debris is swept from the cut; if the gas is oxygen, the endothermic reaction is utilized in the cutting of steel. A gold or stainless steel mirror is shown at A and a lens of germanium at B, with

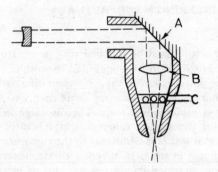

Figure 5.7
*The design of a
nozzle for a low-
power laser*

the gas inlet at C. The lens can be protected by a co-axial gas jet, through which both laser beam and gas escape.

LASER CUTTING MACHINES

The firm of Laser Cutting Limited are pioneers in the utilization of this power source, and the laser now offers many outstanding advantages over traditional cutting systems. Combined with purpose-designed profiling tables under NC or optical control, the laser is now profiling stainless, mild and spring steels, Nimonics, titanium, acrylic plastics such as Perspex and wooden die forms for the packaging industry. The accurately controlled light beam (± 0.015 mm) does not require physical contact with the material and therefore eliminates the problems of clamping very thin gauges even down to 0.05 mm. Similarly high cutting speeds of up to 10 m/min for stainless steel and plastics can be achieved.

The intense cutting power and minimal heat effect of the minute laser beam also ensure a typical heat-affected zone of less than 0.55 mm and a cut width of 0.1 mm. This allows maximum utilization of material while avoiding problems of cutting heat-sensitive metals and plastics. Figure 5.8 illustrates a laser cutting machine in the works of Laser Cutting Ltd; some typical examples of production are now listed.

EXAMPLES
Equipment
The NC 2000 L Laser–Ferranti MR400 laser has a power of 500 Watts and is used under NC or line-following control. The table area is 2 m X 1.6 m and the Cutting speed is 5 m/min.
Applications
Die forms can be laser cut in wood for the packaging industry. Other applications are the medium batch production of valves plates in stainless steel, profiled letter, motifs and graphic designs for the architectural and sign industry and laser profiling hard elastomer polyurethane seals and gaskets.

Figure 5.8
An illustration of a laser cutting machine (by courtesy of Laser Cutting Ltd)

Equipment
The NC 700 HS Ferranti MF400 laser has a power of 500 W and is used under NC. The table area is 7 m X 0.7 m, and the cutting speed is up to 10 m/min.
Applications
Applications include high-speed production of letters and motifs in acrylic plastics for the sign industry, distortion-free cutting of honeycomb and porous steel materials, complex profiling of engineering components in special and tool steels and laser cutting mild steel shim stock down to 0.002 in thick.
Equipment
The LF 3000 Ferranti MF400 laser has a power of 500 W and is used under optical line-following control. The table area is 3 m X 2 m, and the cutting speed is up to 2 m/min.
Applications
Applications include profiling of pipe flanges in stainless steel, shape cutting of architectural designs in coated steels and shape cutting in mild steel, Nimonics or titanium.

CERAMIC SCRIBING

The use of thick film hybrid circuits has led to the production of alumina ceramics in the form of large sheets. Existing methods which use either a diamond tool to scribe the fired material or a system that pre-scores the ceramic, have proved to be both inaccurate and expensive. In contrast the stable and precise scribing beam of the laser ensures a scribe tolerance of ± 0.025 mm when measured between the centres of adjacent scribes.

LASER CUTTING MACHINE CONFIGURATIONS

Several different configerations have been adopted for cutting machines using lasers. One embodies a stationary laser and nozzle, while the work-piece is mounted on a moving table, but the most commonly used type is one with a compact and light-weight laser head that is moved over the workpiece.

The mobile head has the advantage of low sensitivity of vibration when mounted on a rigid structure but, owing to the weight of the laser head, the system has been limited to lasers with outputs up to about 500 W. As a result the thickness of metal cut at the maximum feed rate is limited to about 1 mm for metal cutting. In Germany, however, at Messer Griesheim GmbH, a machine is in operation with a 1000 W carbon dioxide laser with a cutting speed of 8 m/min on structural steel. The mass of the unit has been kept to 300 kg by the use of light-weight metal in its construction.

The laser is suited for use in fine work such as welding fine wires and small mechanical parts encountered in micro-electronics and the watch industry for the weld pool and heat-affected zone is quite localized. A particular advantage of lasers is that welds can be made in otherwise in-acessible locations. For example, mineral-insulated thermocouple wires 25 μm in diameter have been fused inside the 250 μm outer sheath by focusing the laser beam into the ends of the sheath. Lasers of certain types have proved effective for drilling if the hole can be produced by a single shot or small number of pulses. Thus a single pulse can make holes of up to 500 μm in metal sheets up to 500 μm thick. For operations on gem stones for watch bearings many pulses of low power are used to produce the hole.

6

UNIT CONSTRUCTION OF MACHINE TOOLS

The advantages of using unit heads for machine tool construction are that the heads can be assembled around the component to be machined, or they can be arranged in line as in automatic transfer machines. Alternatively, they can be arranged around an indexing table for rotary transfer operations. The angle of approach of the tools to the work can be horizontal, vertical or at any angle required, and the layout can be easily changed to form a larger or smaller machine if required. The simplicity of the units reduces the risk of breakdown, for there are no elaborate gear boxes employed and speed changes, if required, are generally obtained by means of pick-off gears.

There are comprehensive American standards which are divided into single-station or multiple-station machines, these being subdivided into various types and sizes using basic elements. These include main bases, wing bases, feed units, horizontal adapters, and angular and vertical columns. All the standard requirements have been drawn up on the basis of a work-loading height of 42 in. (1066 mm).

FINE BORING MACHINES

Prominent in the use of unit heads is the class of fine boring machine tools whose purpose is the removal of small amounts of metal with resultant great accuracy and a high surface texture. Cutting tools may be of cemented carbide or diamond running at high speed in a boring bar connected to the headstock spindle. This unit is generally mounted on pre-loaded bearings, the accuracy of mounting enabling bores to be produced to size to within 0.005 mm and to 0.002 mm for roundness and parallelism. Machines are of the horizontal construction type with one to, say, six heads arranged to suit the machining operations. A typical machine is that of Figure 6.1 showing a double-ended type for operating on a rear axle gear case. The front spindles finish the bore and face the recesses for the pinion bearings and oil seal, as well as chamfer. The rear tools finish the face and bore the recesses for the differential bearing housings at the second setting of the work. The feed motion is by hydraulic operation, because it has the advantages of stepless rates of traverse, and the feature that the saddles can

Figure 6.1
A double-ended fine boring machine for machining rear axles

be run against dead stops without fear of damage for accurate facing or
recessing depths.

DOUBLE-BANKED HEADS
Figure 6.2 shows an example of a duplex six-spindle fine boring machine,
the two heads on the right-hand side of the machine being mounted above
the others. Unit heads are used in great numbers for machining automobile
components, connecting rods being a common example, and in the case
shown two rods are mounted vertically in the fixture so that the crank
shaft ends and the gudgeon pin holes are machined simultaneously. The
top left-hand view shows the pneumatic clamping method, the rod being
held by adjustable stops against the cap, together with three plungers for
further location, and by a swivelling clamp acting on the small end of the
rod.

Figure 6.3 shows the machine (M.S.T. Torino) with the connecting rods
seen in the central position. The boring heads are driven by a 5 hp motor,
while the table traverse is by a hydraulic piston and cylinder arrangement
giving a steplessly variable feed rate. All slides are controlled through limit
awitches operated by adjustable cams on the slideway. The pneumatic
clamping system is on the right of the machine, assembled on the hydraulic
generator panel which forms a single unit for all control equipment.

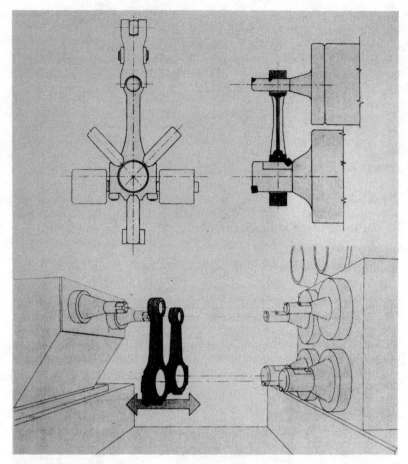

Figure 6.2
An example of a six-spindle fine boring machine with banked heads

ELECTRONIC CONTROL SYSTEM

An electronic control system (Flexcontrol patent) composed of a static logic group on a small scale is located on the machine and replaces any relay cabinets. It has extreme flexibility of utilization, being able to switch from one working cycle to a completely different cycle with the simple replacement of a programme card in the form of a diode circuit inserted in the base module through a multiple plug. The system drives static power relays for the control of 110 V a.c. equipment such as electro-valves or remote control switches.

There is high control reliability and flexibility, an increased efficiency of the machine with easy maintenance resulting from the modular construction, and an indication of the operative condition of the machine by means of light-emitting diodes. Thus neither highly skilled personnel nor specific programming knowledge is required.

The push button control stations located on the front of the machine cycle to be controlled. These cycles include an automatic operation, a semi-automatic operation with sequential working by push button either for idle or working trials or thirdly a cycle working manual independent controls for single movements.

The numbers of unit heads may range from one to six, with a spindle speed of 6000 rev/min.

TYPES OF UNIT HEADS

MECHANICAL HEADS
The heads shown in Figure 6.4(a) and Figure 6.4(b) can have spindle-driving motors ranging from 1½ to 20 hp to suit four sizes of heads. For

Figure 6.3
An illustration of the machine for boring connecting rods
(by courtesy of MST Torino)

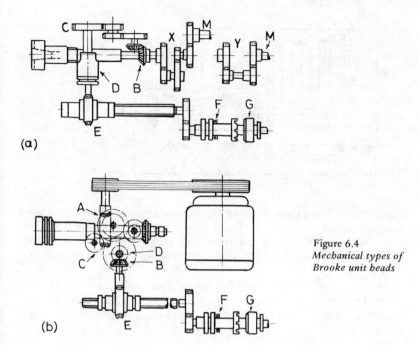

(a)

(b)

Figure 6.4
*Mechanical types of
Brooke unit heads*

the two smaller units shown in Figure 6.4(a), the drive is from a flanged
motor to shaft M, and thence by pick-off gears, as at X, using six gears to
give a low speed or, as at Y, to give a medium- or high-speed drive. The
two speed ranges are 103 to 3000 or 85 to 2000 rev/min respectively. The
larger heads shown in Figure 6.4(b) incorporate built-in stator-rotor units,
and the drive to the spindle is by vee belts and worm and wheel A. The
pulleys can be easily changed to vary the speed of the drive, and the stator-
rotor housing is eccentrically mounted in the casing to enable the tension
of the belts to be adjusted. On the smaller heads of this type, the two
speed ranges are from 65 to 300 or from 300 to 1500 rev/min, and on the
larger head from 36 to 122 or from 182 to 925 rev/min.

The motor sizes given provide drilling capacities in cast iron for
diameters of $1\frac{1}{8}$, $1\frac{3}{4}$ $2\frac{1}{2}$, and 4 in. The spindles mounted in roller bearings,
have standard milling machine tapers conforming to *B.S. 739.* Either right-
or left-hand rotation is available. On the designs in both Figure 6.4(a) and
Figure 6.4(b) the feed is obtained from the main spindle by bevel gears B
and pick-off gears C through a multi-disc overload clutch D and worm
wheel E keyed to the feed nut. The lead screw mounted in the slide base
of the unit is prevented from rotating by a multi-disc brake F during the
cutting stroke. The fast traverse is at the rate of 130 in./min, and overruns
the feed motion; it is derived from a separate motor of ½ to 1½ hp according to the size of the head. The motor is flanged mounted at the rear of
the slide base and drives the lead screw through a multi-disc overload

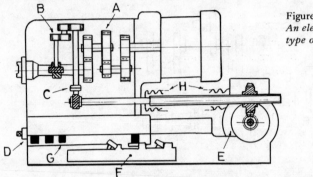

Figure 6.5
An electro-mechanical type of unit head

clutch G and a gear train. The lead screw brake is automatically disengaged as soon as the motor is switched on by the action of steel balls which move over cam faces. The length of traverse may range from 8 to 30 in. according to the size of the head which can be mounted horizontally, vertically or at any angle.

The automatic cycle is controlled by adjustable dogs carried on the side of the head; these operate limit switches and the spindles rotate throughout the cycle. Coolant only flows when the head is running. Milling heads fitted with spindles at right angles to the slideways are available.

ELECTRO-MECHANICAL HEADS

Figure 6.5 shows a design in which the spindle is driven through gear trains at A, and a spiral gear on the spindle drives the vertical shaft to engage the feed train B. This continues to another spiral gear mounted on the lead screw. A slipping clutch is fitted into the feed drive at C, so that in the event of overload it will disengage the drive and prevent damage. It also enables the head to run against the end stop D for facing operations, thus ensuring accurate work length by slipping of the clutch.

A fast feed motor is fitted at E, but normally the direction of rotation of the worm wheel enables the screw to wind itself out of the nut to give the slow movement for the cutting action. When the feed motor revolves, the drive by worm and wheel rotates the nut at a higher speed than that

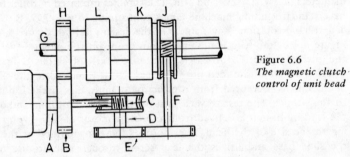

Figure 6.6
The magnetic clutch control of unit head

obtained by the normal gear train. This gives a differential movement in which two velocities are added during the fast approach and are subtracted to give the rapid return of the head. The feed motor is fitted with a magnetic brake to ensure rapid stopping of the traverse. Changes from forward to reverse motion as well as variations from fast to slow rate are accomplished by limit switches mounted in the box F on the side of the base. The switches are contacted by adjustable dogs G which can be set to give the cycle and length of traverse required.

MAGNETIC CLUTCH OPERATION

Figure 6.6 shows a unit in which the drive is by the electric motor on shaft A carrying wheel B and worm C meshing with the worm wheel on shaft D. Pick-off gears at E connect to shaft F. Parallel to shaft A is shaft G coupled to the feed screw. On shaft G is wheel H and worm wheel J, both free running, but keyed on G are the coil parts of magnetic clutches K and L with portions of these clutches bolted to the worm wheel and gear, respectively. For the feed motion, current is supplied to clutch K only, and the coil in the clutch L is de-energized. Thus the drive is transmitted from shaft A through the worm and wheel and change gears to shaft F.

With the clutch K energized, the wheel J is coupled to shaft G and drives it. The gear B on the shaft is also being driven, as is the gear H with which it meshes. Since the clutch L is de-energized, however, the gear H runs free on shaft G which is slowly rotating for the feed motion. For the rapid traverse motion, the clutch L is energized and clutch K de-energized, so that shaft G is driven by gears H and B at a fast rate, since no reduction is involved. The feed section is still being driven during the rapid traverse but with the clutch K released and the worm wheel rotating freely. The clutches are operated by limit switches actuated by trip dogs on the machine table or slide, and the advantage being that magnetic clutch operation is practically instantaneous.

CAM-TYPE UNIT HEADS

The design shown in Figure 6.7 is simple, requiring one motor drive only

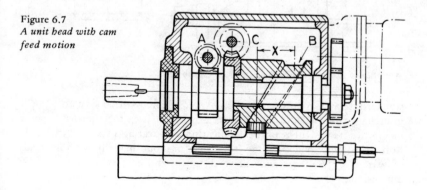

Figure 6.7
*A unit head with cam
feed motion*

and no electrical switches. The limitation is the length of the traverse available, this being controlled by the contour of the cam B on which X denotes the traverse. The rate and period of feeding during a cam revolution can be varied to give rapid approach, feed forward for cutting, rapid return and a period of dwell for work changing. All these can be varied by the contour of the cam but, without varying the change gears for driving the cam, they can only be varied by changing the cam. The position of the traverse of the saddle relative to the workpiece can, however, be varied by adjustment of the nut by the screw in the slideway.

The drive to the spindle is by a flanged motor and gearing. A spiral pinion on the spindle drives another spiral gear A mounted co-axially with a pick-off gear located on the outside of the box and meshing with a spur gear C. The feed motion is transmitted back inside the box and terminates with a worm meshing with the worm wheel integral with the drum cam. A different feed rate is thus available by changing the pick-off gears, the actual feed movement being by the rotation of the cam and the ball bearing roller engaging the cam profile.

TOOLING EQUIPMENT FOR UNIT HEADS

One of the advantages of using unit heads for batch or mass production is the facility with which tooling can be carried out by attaching multiple spindle units to the front of the heads. These attachments can be used for multiple drilling, boring or milling operations and can operate simultaneously from several directions if required. In view of the large number of heads being built, the British Standards Institution (B.S.I.) have published a specification *B S 3295* for heads of the slide type (Figure 6.8). The

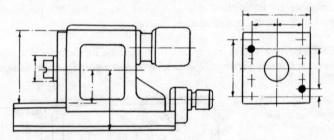

Figure 6.8
A unit head showing dimensions standardized by BSI

standard relates to those dimensions and elements necessary to achieve interchangeability both of mounting and of use, together with some qualitative terms such as the application of locating pins, shown black. Bolt holes are provided on the front face for mounting multi-spindle heads of various types; the standardized distances and heights are indicated by dimension lines.

The size of the unit heads are related to the power of the driving motors, these being specified in *BS 2960*, and are grouped as follows: (a) 1 to 3 hp; (b) 5 to 7.5 hp; (c) 10 to 15 hp; (d) 20 to 25 hp. The dimensions of the spindle noses are those of *BS 1660*, Part 3, except for the dimension from the spindle end to the face of the unit, these being given in the specification. Six bolt holes are provided on the front face for units up to 7½ hp, and eight for the larger units.

The units are designed for direct tooling in the spindles if desired, but there is no direct relationship between the horsepower and the thrust for the variable volume of metal removed in a given time, so that a positive value cannot be given. As a guide to expected performance, the minimum axial thrust may be assessed at 1000 lb per unit horsepower of the driving motor.

MACHINING HEADSTOCK BORES

An example of the use of multiple spindle attachments applied to unit heads is given in Figure 6.9. This shows an end view of the boring jig of

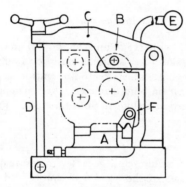

Figure 6.9
A boring jig for holding headstocks

one of the Harrison lathes, the location of the base of the headstock being by a vee and flat support on the block A. This is removable and can be replaced by another block to accommodate other sizes of headstocks. The casting is held down on its location by the swinging clamp B in the lever C, clamping being by means of the swivel bolt and the handwheel. To facilitate insertion and removal of the castings, bolt D, the lever is fitted with a counterweight E which brings the lever clear of the casting as soon as the handwheel is released. End location of the headstock is by the screw F, forcing it against a dead stop.

The three-spindle unit detailed in one plane in Figure 6.10 is also shown with the centres in the correct position in the end view. The drive from the unit head comes to the slotted end of shaft A which runs at 337 rev/min and connects to the main spindle B which runs at 207 rev/min. The remaining gear transmission gives speeds of 438 rev/min to the other two spindles. A cutting speed of 207 ft/min (63 m/min) on the main headstock bore

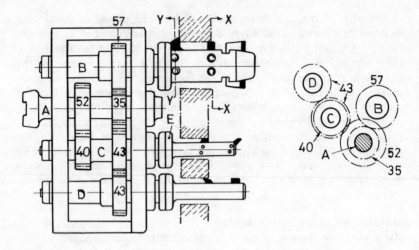

Figure 6.10
A tooling arrangement for boring headstock
(by courtesy of T S Harrison & Son Ltd, Heckmondsike)

100 mm in diameter is used, with 215 ft/min (66 m/min) on the other two gear shaft bores. Cutting takes place simultaneously on both ends of the headstock, the tooling being similar on both ends except for bore diameters. For these three main bore diameters, the roughing and finishing operations are performed at a floor-to-floor time of 23 min per headstock.

CUTTING TOOLS

The three centres indicated require close accuracy to ensure quiet running of the gears. Thus boring bars are short and rigid, and instead of mounting them in taper bore spindles, which necessitate considerable overhang, each bar is directly coupled and located to the driving shaft by a flange as at E. The heavy bar for machining the main bore for the lathe spindle bearing carries three cutters, a front double roughing cutter, followed by a sizing cutter ½ in. in round section and finally a recessing cutter. The sizing cutter is carbide tipped and adjustable for size before final clamping by the two screws in the bar.

Figure 6.11 shows a section through the bar at X-X, while a section through Y-Y shows the holes for the recessing cutter, some of the machining coming into the flange. The centre of each of these cutters is offset 0.14 in. from the centre of the bar, enabling this to be kept as large as possible in diameter and still enabling the cutting edge of the tool to extend sufficiently for cutting.

The large double-ended roughing cutter is shown located on two flats on the bar and clamped by a screw in the end of the bar but, for the opposite end of the headstock spindle bore, a carbide-tipped cutter is used.

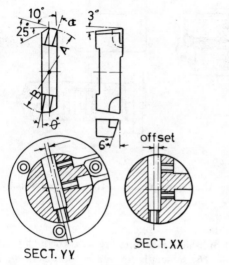

SECT. YY

SECT. XX

Figure 6.11
A detail of boring cutters

A feature is that one cutting edge precedes the other by 0.1 in. and each cuts a different diameter. For example, while diameter A is 3.781, B cuts 3.531 in., the angle α being 9° and θ equal to 10°. The advantage is that B acts as a semi-finishing cutter and leaves little metal to be removed by the sizing edge of A.

The construction of the unit-head machines comprises only a simple bed structure, the unit heads, a plain table for mounting the fixtures and the attachment carrying the tools. Yet the versatility of the machines and the high output obtained far outpasses that of several standard machine tools with their complexity of driving and feed boxes, many of the speeds and feeds never being required for batch production, and simply adding to the initial cost of the machines.

ELIMINATION OF VIBRATION BY CORRECTIVE DESIGN

For machining long bores an outboard support may be used, but there are some disadvantages. They can only be used for through holes, and the bar must be far longer than for an unsupported one. If the bar is of sufficient diameter, it is an advantage to reduce the end portion so that it may revolve on a ball bearing support. The importance of this feature is that, whereas a bar 25 mm in diameter running at 600 rev/min will have a surface speed of 48 m/min, a 100 mm bar running at the same speed reaches 160 m/min. It is possible, however, to support a bar close to the cutter, as shown in Figure 6.12(a). The support is part of the headstock and comprises a steel sleeve A flange mounted on the headstock which carries the revolving spindle B supported by the pre-loaded bearings at C. The outside diameter of the sleeve must be slightly less than the machined bore produced by the cutter head D at the end of the bar. This is an effective method of reducing vibration for the tool is well supported at all times,

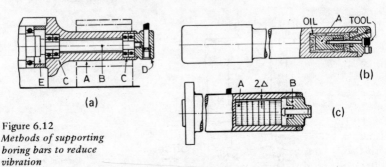

Figure 6.12
*Methods of supporting
boring bars to reduce
vibration*

while the drive from the main spindle to that of the bar is by the flexible coupling E.

VIBRATION DAMPING

A large number of unit-head machines are used for high-precision boring using diamond tools. The subject is covered in *Manufacturing Technology*, Volume 1, *Basic Machines and Processes*, but means of damping vibrations are indicated by the diagrams in Figure 6.12(b) and Figure 6.12(c). The rigidity of a bar can be increased by using a material with a higher modulus of elasticity than that of normal steel. The rigidity is directly proportional to the modulus and the fourth power of its diameter and inversely proportional to the third power of its overhung length. Length-to-diameter ratios of 4 to 1 are suitable for steel but may increase to 7 to 1 for bars of tungsten carbide.

A high-strength powder metal composition developed for tool support is known as 'Mallory-No-Chat' (Johnson Matthey Metals Ltd). It incorporates tungsten and has a density of three times that of steel, a modulus of elasticity in tension of 40×10^6 lbf/in.^{2}, and a tensile strength of 50 tonsf/in.^{2}. A tool shank of this material will thus deflect less under a given load than a similar steel shank and is suitable with brazed tools for boring or turning.

Figure 6.12(b) shows a tuned and damped absorber where, by suitable tuning of the frequency and adjustment of the dashpot, smooth cutting and good surface finish on the workpiece can be obtained, even on bars with a length-to-diameter ratio of 10 to 1. The damper consists of a tungsten-based alloy slug A in an oil reservoir, so that under cut the bar tends to vibrate but the slug tends to remain stationary; the cushion of oil is pushed around as relative movement occurs to dissipate energy in the form of viscous friction.

Figure 6.12(c) shows a multi-mass vibration damper which operates by 'soaking up' vibrations in a boring bar. There are inertia discs A located in a row, the discs being of slightly varying diameters. Any vibration causes the discs to bounce so that they hit the inside of the bar in random timing and therefore reduce the vibration below the amplitude that will cause chatter. On average the bar still moves at its natural frequency, but amplitude may be reduced by as much as 90%. The discs are compressed axially by a spring in the adjustable sleeve B, the compressive force determining the stiffness and non-elastic resistance forces in the joints between the discs which may be of steel or heavy alloy. From tests on various bar diameters and numbers of discs the optimum values proved to be as follows: diametral clearance $2\Delta = 3$ mm; pressure on the discs within 1 to 2 kgf. The best results were obtained with eight discs. The boring bar diameters varied from 15 to 70 mm and, when diamond boring bronze with the 15 mm bar, the limiting value of the length-to-diameter ratio was 7; it equalled 8 for the 70 mm bar.

Where high precision is required from a boring bar mounted in a unit head and where the bore length may be extreme with respect to its diameter, plain bearing supports are often essential. In this case pressure lubrication often by oil at 50 lbf/in.[2] is employed for the outrigger bearing, and the casting supporting that bearing is frequently cooled by the liquid being pumped through the casting and around the bearing. Figure 6.13 shows a special purpose machine for boring the crank shaft

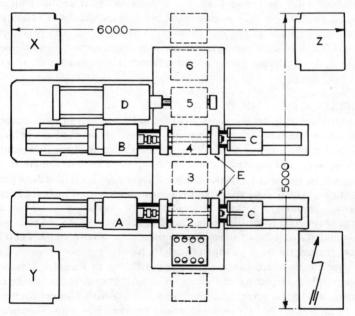

Figure 6.13
A unit head machine for diesel engine cylinder blocks

and cam shaft bores of a vee-8 diesel engine. The bore is 2.5 in. in diameter for the cam shaft bearings and 27 in. long; no intermediate support bearings can be fitted for the boring bar.

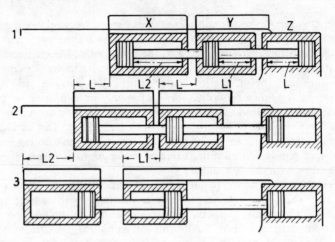

Figure 6.14
The three-position control of a boring machine table

The machine comprises the main unit heads A, B and D and two opposed heads C with the water-cooled bearings indicated at E. The starting position for machining the cylinders is at station 1, and they travel across the machine to the different machining and inspection stations to station 6. The hydraulic power units are shown at X, Y and Z.

HYDRAULIC CONTROL OF UNIT HEADS

An advantage of hydraulic control is that slides can be run against a dead stop without fear of damage, and this feature can be used for accurate positioning as in Figure 6.14. The system shown is for a fine boring machine with tandem tables, these requiring accurate location for facing operation or boring to a precise depth. The design is for three positions but can be extended to six if required. The two slides are caused to move by equal amounts for the first stage, and by unequal amounts for the second stage. The front slide is indicated at X and the rear slide at Y, with the cylinder Z attached to the bottom slide. The distances L, L_1, and L_2 denote the lengths through which the slides travel.

A system in which the table indexes is shown in Figure 6.15(a) and Figure 6.15(b) depicting the electro-hydraulic system for a fine boring machine used for machining a fuel pump housing with 40 bore diameters and 14 faces, the floor-to-floor production time being 3½ min. The slides 1 and 3 are for machining six cylinder bores while 2 and 4 machine two cam shaft and two shaft bores. For the push button arrangement shown in Figure 6.15(a) there is a main starting panel (bottom left), but the machine can be stopped at any of the other three positions. Also any slide can be reset, inch any spindles or index the cross slide from four positions.

The close spacing of the bores prevents all six from being machined at

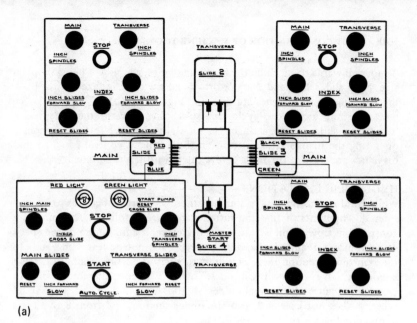

(a)

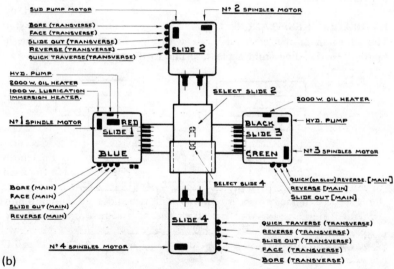

(b)

Figure 6.15
A hydro-electric control system for a fine boring machine

one time so the table is indexed transversely. The cycle is as follows. First slides 1 and 3 move inwards to machine the first, third and fifth cylinder bores from both sides while slide 4 moves inwards to bore at one end, slide 2 being inoperative. Then slides 1, 3 and 4 are returned to the starting position. The work table is indexed to the second position to bore the second, fourth and sixth holes as slides 1 and 3 again move inwards. Slide

2 moves inwards to machine the remaining holes, the slide 4 being inoperative. Now slides 1, 3 and 2 are returned and the table indexed to the loading position. Limit switches are shown in Figure 6.15(b).

Before the automatic cycle will start, the operator must depress a button on slide 4 with a finger on his right hand while simultaneously operating the main starting button on the left-hand control panel so that his hands are clear of the tools before the inward travel commences.

HYDRAULIC LIMITATIONS

While fluid operation is very flexible, there are certain features to be considered. The limits of fine and fast traverses are as follows. The minimum controllable flow of oil is 5 in.^{3}/min; therefore, if a minimum piston speed of 1 in./min is required, the cylinder area must not be less than 5 in.^{2}. In the opposite direction there is a limit to suction oil flow (about 20 ft/s), beyond which suction oil will not follow up a circuit, thereby tending to create a vacuum. In general, pipes should have a diameter so that oil velocity does not exceed 2 to 4 ft/s under suction, and from 8 to 15 ft/s in pressure lines.

HYDRAULIC RESISTANCE

The adoption of the 'electrical' view with regard to control is shown in figure 6.16. Leading from a pressure line to the reservoir, a line may be

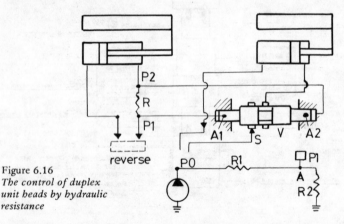

Figure 6.16
The control of duplex unit heads by hydraulic resistance

taken from A to obtain a pressure P_1 which will be a fixed fraction of the input pressure P_0. If these resistances are designated R_1 and R_2, then the pressure at point A will be

$$P_1 = P_0 \ \frac{R_2}{R_1 + R_2}$$

An example is shown in which on an automatic machine one slide comes to rest before movement of the second one commences. The problem may be that there may be no moving member from which the control valve of the second slide can be operated. In the diagram the resistance coil R is placed in series with the hydraulic line leading to the cylinder of the first slide. The control valve V for the second cylinder is coupled in parallel with the resistance coil and is operated in opposite directions by two plungers of unequal areas A_1 and A_2. While the first slide is in motion, oil is flowing through the resistance R, thus creating a pressure difference, P_1-P_2. When the slide reaches the end of its stroke, the flow through the resistance R stops and pressures P_1 and P_2 become equal. The values of the resistance R and the areas A_1 and A_2 are such that, when the slide is moving, the product of the pressure P_1 and the small area A_1 will be greater than the product of the pressure P_2 and the large area A_2. Thus, the control valve V is held in its extreme right-hand position. However, when the slide has completed its movement, and P_1 is equal to P_2, the valve moves to the left, so permitting oil to pass from the supply line S to the cylinder of the second slide.

BALANCED CIRCUITS

Balancing valves can be used for speed control under varying loads, the Wheatstone bridge being the nearest electrical analogy. From the diagram of Figure 6.17, if pressures at E_1 and E_2 are equal, the flow through the

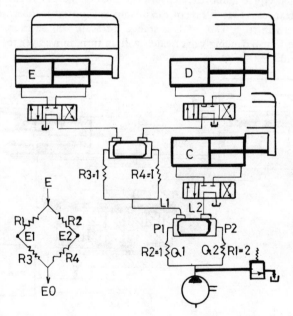

Figure 6.17
A diagram showing the use of balancing valves for headstock control

resistances R_1 and R_2 will be inversely proportional to those resistances. If P_2 exceeds P_1, the valve will move to the left, decreasing the resistance R_4 and increasing R_3. This causes P_1 to approach and then to equal P_2. The flow in each side of the bridge is thus

$$P_0 - P_1 = Q_1 R_1, \quad P_0 - P_2 = Q_2 R_2$$

Since the valve causes P_1 to equal P_2 at all times, the right-hand side can be equated. The ratio of flow will therefore be

$$\frac{Q_1}{Q_2} = \frac{R_2}{R_1}$$

Q_1 and Q_2 are independent of pressure in the system. If a constant quantity Q be supplied at P_0, the flow in each branch will be in the ratio determined by the formula. If the resistances R_1 and R_2 are equal, the flow in each side of the bridge will be equal regardless of the loads L_1 and L_2, either of which may be zero. The balancing valve thus provides another means by which the several slides can have movement in a required relationship.

The diagram shows three pistons of equal areas, all of which will move at the same rate, irrespective of load variations if the resistances are in the ratios shown. If R_1 is twice R_2, then from the equation given there will be exactly twice the flow through R_2 as through R_1. The flow through R_2 is then divided equally, and the deliveries to the cylinders C, D and E will be equal. A feature is that, if one of the pistons encounters an obstruction, all the pistons will stop. Thus, if C is stalled, pressure will be exerted on the left of the first balancing valve which will, in turn, move over to interrupt oil to D and E.

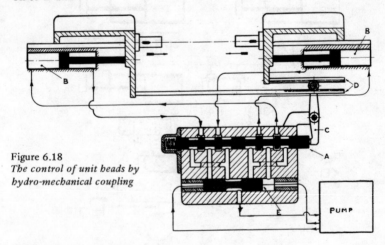

Figure 6.18
The control of unit heads by hydro-mechanical coupling

MECHANICAL COUPLING
To ensure that the saddles of a duplex unit head machine move at the

same speed, a valve A (Figure 6.18) is arranged to control the oil flow to cylinders B. The valve is moved in one direction by a spring and in the other by lever C which is actuated by the racks D and pinion. No flow of oil to valve A can take place until reverse valve E is located at either end. It will be apparent that, so long as the speed of travel of the two saddles is equal, the pinion on lever C will be merely rotated without affecting the relationship of the two saddles but, should one saddle begin to slow up, lever C will move from its vertical position and the oil supply to the leading sadddle will be restricted until an equal feeding movement to both saddles is obtained. Without this feature: the cylinder encountering the least resistance may take the entire oil flow until its stroke is finished. This may well take place in a drilling machine, for example, where different-sized drills are being used.

MULTIPLE DRILLING HEADS

In view of the great number of holes found in engineering components, drilling forms a major operation often with unit heads as shown in Figure 6.19 which illustrates a machine with two opposed horizontal heads and

Figure 6.19
A machine for drilling back axle castings

one vertical operating on automobile back axle casings. The feed traverses are by hydraulic operation from the pump unit seen on the left, with all driving and feed motions controlled by push buttons which allow for a full automatic cycle or independent control.

So great has been the development of transfer machines with unit heads, that standard drill spindles are available to suit any hole pattern. Dex Gears Ltd, Coventry, specialize in unit-head drilling spindles, and more than 1,000 different spindle–bearing combinations are available.

For drilling small and large concentration of holes at close centre distances, gearless heads are available and will drill as many as 60 holes in component areas of only 9 in.^{2}, or spindles can operate at 7 mm (0.27 in.) centres. The drive to the spindles is transmitted by an orbiting plate and small throw cranks. Zager-Dex can supply spindles at centres down to $^1/_8$ in. and have made a unit head with 500 spindles.

D.C. DRIVE UNIT HEADS

Some of the unit heads described have complicated drive mechanisms using up to two motors for feeds and rapid traverses and one spindle motor. These fixed-speed a.c. motors require a large number of gears, shafts and sometimes clutches to provide the necessary movements. Unit heads are also available which use a single d.c. variable-speed motor which is able to provide both feed and rapid actions without more than one pair of gears and a lead screw. In the same way, the spindle is driven from a d.c. motor which is able to provide a wide range of speeds without having to resort to pick-off gears. Two spindle speeds and two feeds are sometimes used on the same drilling operation where a stepped drill must be used. Cutting speed and feed for each diameter of the drill is thus optimized, giving the minimum operation time with the maximum tool life.

The cost of such a unit is very little more than some of the more complicated ones described earlier in this chapter. Mechanical parts are fewer with a high degree of consequent reliability. Micro-switches and cams are used to change the speeds and feeds at the appropriate point in the travel of the unit, and one manufacturer has even included a simple decade switched NC to permit flexible unit-head machining. A description of decade switch programming is given in Chapter 10.

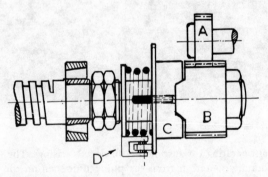

Figure 6.20
Slipping clutch safety device

FEED PROTECTION MECHANISM

Where a unit head has a mechanical feed traverse, a safety device can be fitted to the lead screw. Figure 6.20 shows an arrangement in which the driving pinion A, coupled to the motor drive, meshes with the wheel B which runs freely on the feed screw. The wheel B has tapered clutch teeth to engage similar teeth in unit C which is keyed to the feed screw. The units B and C are normally kept in engagement by the pressure spring while machining proceeds but, if the saddle contacts a stop, the clutch teeth slip out of engagement. Normally, this provides the safety device but, in the arrangement shown, there is an added refinement of a micro-switch D which operates when the unit C disengages from B, the action not only stopping the forward feed movement, but reversing the direction of rotation of the feed motor and hence the traverse of the head.

7

DEVELOPMENT OF AUTOMATIC TRANSFER LINES

PALLET TRANSFER SYSTEMS

The method of transportation of work which is not circular is on pallets with clamped components, so that transfer is by pawls or rotating bars with cams, which, in the operating position, enter slots in the base of the pallets to prevent over-running at the end of the bar movement. The pallets return to the starting position after the machining operation. In some cases they are returned on a conveyor placed directly under the work fixtures. The pallets are lowered to this position at the end of the machining line, but the drawback is the limited height between the floor and machining operations, and the fact that swarf collectors and conveyors are difficult to install. Also the unit heads and inspection stations, being attached to the transfer line, cause the formation of a tunnel under the line and make the inspection of the conveyor and cleaning of it difficult of access.

It is preferable then that the pallets return over the machines as shown in Figure 7.1. Another advantage, in addition to accessibility, is that the

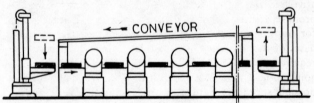

Figure 7.1
The transfer of pallets with an overead conveyor

return conveyor may be of a simple roller type on which components return under their own weight or with little effort, so saving a power unit. This system is not feasible if vertical machines are mounted in the line, and braking of the components may bave to be considered if the angle of the conveyor is steep.

An alternative method is shown in Figure 7.2. The pallets are moved horizontally only and are returned by a conveyor placed on the side of the line at the same height as the main conveyor which carries the work through the machines. The main disadvantage is the increased floor space

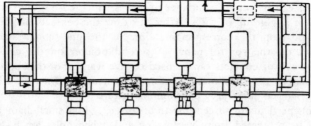

Figure 7.2
A pallet movement side conveyor system

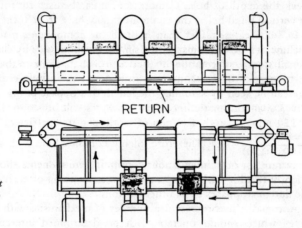

Figure 7.3
*An arrangement
of conveyors to
save space*

required by the machine and, if this feature is of importance, an alternative but more complicated system is available (Figure 7.3). This is a compromise, for the pallets are first displaced horizontally and then vertically for transfer to the returning conveyor. The latter passes through the side bases of the machines and does not cause an increase in the width of space required. When the work is tall, this method may not be convenient if it is going to lead to weakening of the side bases of the unit heads. The highest possible speed of a return conveyor should be used to keep the number of pallets in the line to a minimum. This reduces the number of pallets on idle stations and on the return conveyor.

EXAMPLES OF OUTPUT

Two cases will be considered. One is the machining of cylinder heads for vee-type engines, a line of 60 machines producing 170 heads/h when served by two operators. The other is vee-cylinder blocks producing 85 blocks by 85 machine tools, for again by two operators. Stoppages owing to tool changing are inevitable, but the time can be reduced by using quick-acting tool-changing devices. In some cases 1,000 tools may be in the line, the majority being non-precision tools. These can be divided into groups with about the same tool life, and a cycle counter can be fitted into the line and can be set to a number of cycles which corresponds to the life period of any group of tools. After this time the contacts on the counter close and stop the corresponding flow of the line, and a signal lamp operates on the control panel. A second pointer on the counter and another pair of contacts can be used to give warning of the approach of a tool change, this preliminary warning being necessary for tools which cannot be changed quickly, eg. face milling cutters.

Tools particularly susceptible to breakage, eg long small-diameter drills, should be protected from overloading and, when holes are machined in sequence by a number of tools, inspection devices should be fitted to check the depth of hole. Compressed air can be used for this purpose but, for vertical blind holes, the component may have to be turned over.

In some cases, rather than have all the components machined on one machine, it is better to install two similar machines side by side. This allows a small storage of components between each machine so that, in the event of stoppage of one machine, the flow does not entirely cease. The advantage can be seen by assuming the length of stoppage of one machine to be only 3%; an automatic line of 50 machines will be stopped on an average of 150 min for each 100 min of work, and its coefficient of use will be $\frac{100}{(100 + 150)}$ = 0.4. Therefore, highly complex lines with a high productivity rate for machining labour-consuming components should be of sufficiently flexible design to prevent excess stopping of each section of the line due to stoppages of its other sections.

Automatic lines represent a number of mechanisms with self-contained drives which should operate with pre-determined interconnection and sequence. When separate electric motors are used for each of the independent movements, control by direct energizing and switching-off motors leads to each motor in the line working for an insignificant part of the time but needing to be switched on and off for each working cycle, causing expenditure of energy for starting. Nevertheless, electrical systems working on high currents are now so reliable that hydraulic operation is being replaced in many cases.

HYDRAULIC OPERATION OF TRANSFER SYSTEMS

A reduction in the energy requirements is of importance to increase the reliability of the control circuits. This can be achieved by means of series application of the servo principle using pilot valves of small diameter and traverse for the control of the hydraulic movements. These allow the pulling force of electro-magnets to be as low as 3½ lbf (1.5 kgf), reducing the power required for the control circuits to that used for low-power apparatus. This feature not only increases the reliability of the control systems but also reduces the size of the control cabinets and panels.

The available hydraulic power can also be used for clamping workpieces onto the fixtures and fixtures onto the pallets. This is in addition to the auxiliary functions of work transference, elevating, lowering or indexing of work. The economy is also increased if several components of similar types can be machined without serious alterations to the unit heads.

CLAMPING DEVICES

Clamping devices often include locating plungers followed by direct-acting clamps or wedge-clamping means for holding components and pallets. The clamping cylinders are generally controlled from one distributor valve, in which case the pistons do not move simultaneously but in order determined by the friction forces in the cylinders and driven mechanism. The working pressure for clamping in a transfer line can be obtained from the double pumps working at pressures ranging from 300 to 700 lbf/in.² (21 to 50 kgf/cm²) for the first pump and from 700 to 1100 lbf/in.² (50 to 76 kgf/cm²) for the second. In a direct clamping system the same pressure can be used for clamping and release where the friction forces are greater. The clamping pressure can be reduced if clamping takes place by oil entering the piston rod end of the cylinder with its reduced area.

Locating plungers should not operate at more than 250 lbf/in.² (17.7 kgf/cm²) to prevent damage in case of misalignment. After location the clamps can be actuated by a hydraulic signal, caused by an increase in pressure in the locating cylinders, but a better means is to actuate the clamps by an electrical command caused by a trip switch controlling the plunger movement. When the clamping is carried out in several stages by consecutive engagement of the clamping cylinders, their engagement should be controlled by electrical commands signalled by the ending of the previous operations. The control of direct clamps and wedges can be by hydraulic pressure relays, operating according to the increase in pressure in the clamping system. The pressure relay can also be used for clamp withdrawal, but with wedge clamps the release can only be controlled reliably by electrical terminal switches.

The number of locating cylinders working simultaneously in one section is usually not more than 15, and the clamping cylinders 25. Thus the capacity of the pumps can be obtained from the number and diameter of the cylinders and strokes of the pistons.

CYLINDER CALCULATIONS

As an example of calculation, taking a single cylinder, having determined the maximum load for the power stroke, this load, divided by the maximum pump pressure, determines the area of the cylinder. The pump delivery is obtained by multiplying the area of the cylinder by the maximum speed, and the delivery so obtained determines the pump size. Thus, assuming the force necessary to be 3000 lbf and the pump pressure to be 300 lbf/in.², then 3000/300 = 10 in., or a cylinder diameter of $3\frac{5}{8}$ in. (92 mm). The pump delivery equals that area of cylinder multiplied by the maximum speed; assuming this to be 30 ft/min, then pump delivery equals

$10 \times 360 = 3600$ in.^{3}/min or, as 1 gal = 277.3 in.^{3}, pump delivery = 13 gal/min (59 litres/min).

In actual practice the cylinders for the plungers range from 1¾ to 2½ in. (from 44 to 63 mm) in diameter with piston strokes of 2 to 4 in. (51 to 102 mm). Clamping cylinder diameters range from 1¾ to 5 in. (from 44 to 127 mm), and in direct clamping systems the piston stroke is about ½ in. (12.7 mm) with 2 to 4 in. (50 to 100 mm) for wedge clamping. A discharge rate for the low-pressure pumps serving the locating and clamping mechanism is 2½ to 12 gal/min (11 to 54 litres/min), and for the high-pressure pumps 0.6 to 1¾ gal/min (2.7 to 8 litres/min) is sufficient to obtain a constant pressure for the clamping system.

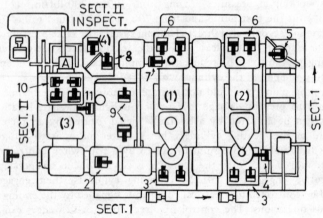

Figure 7.4
A diagram showing the analysis in setting out transfer lines

PROCEDURE IN DEVELOPMENT OF TRANSFER LINES

An analysis of the sequence in setting out a line is illustrated in Figure 7.4. The component is a flywheel housing for a Russian automobile engine. The line comprises two L-shaped sections forming a closed rectangle in which the loading and unloading operations are carried out at one common station. The line includes two duplex vertical drilling and boring machines 1 and 2, a vertical tapping machine 3 and an inspection device 4. Hydraulic power is used on four heads, two of which are for boring, and for operating and releasing power clamps for work holding on the pallets. It is also used for rotating the conveyor bars and moving the pallets in the line sections, for locating and clamping the components at the working and inspection stations and for operating the inspection device for checking the presence and depth of holes for tapping and the electrical control device of the line.

The initial data requirements for setting out a line are as follows:

(1) A diagram of the lay-out of the cylinders and other hydraulic

equipment.

(2) Piston diameters, travel, number of simultaneously working cylinders and their pressure.

(3) A time graph of the working cycle of the line.

(4) A list of special requirements of the hydraulic system, if required.

The line to be described has the special requirements of (a) separate control of conveyors in each section in case of resetting or adjustment of the line and (b) clamping pressure on machine 3 should not be less than 70 lbf/in.^{2} (4.9 kgf/cm^2) with a component clamping control separate from the control used on machines 1 and 2.

At the first stage the total capacity of the cylinders, oil delivery rate and other data are summarized, and the allowable pressures, pump capacities and speeds of travel are settled. At the next stage it can be settled whether mechanisms require braking at the end of a traverse, and the methods can be decided. The methods and devices are also selected for limiting the traverse of the mechanisms, and   finally any variants for special requirements are decided.

From two variants, the first system consists of independent drives, each controlling one or a group of cylinders of one line unit. Four such drives can be used.

(1) Hydraulic operation of the locator, clamping and inspection mechanism on machines 1 and 2, and inspection station. The drive comprises a double pump, valve panels for the locator mechanism, panel for the inspection device and two pressure relays for the clamping and release mechanism.

(2) A hydraulic drive for the conveyor of the first section with a double pump controlling the movements of the conveyor and carrier bars, and throttle.

(3) A similar drive for the second section of the line.

(4) A hydraulic drive for machine 3, for operating the locator and clamping mechanism, and the clamping mechanism on the loading station. This comprises a pump with a capacity of 1¾ gal/min (8 litres/min), a hydraulic valve panel with one safety valve, another safety valve and a pressure relay.

DESIGN ARRANGEMENT

The cylinders in the line are numbered as follows: 1, clamp-operating mechanism, 2, first section conveyor; 3, clamps on machines 1 and 2 in first section; 4, location on machines 1 and 2; 5, second section conveyor; 6, clamps on machines 1 and 2 in second section; 7, location on machines 1 and 2 in second section; 8, control; 9, conveyor bar returning mechanism; 10, clamp on machine 3; 11, location on machine 3.

The second variant can use a centralized hydraulic system in which one double pump serves all the units in the line. This system can only be applied where the mechanisms requiring a large quantity of oil are working successively. A pump of 7¾ gal/min (34 litres/min) is satisfactory; for 67% of the time of one cycle the auxiliary equipment is stationary and during this period it is only necessary to maintain pressure in the clamping system.

Figure 7.5 shows the lay-out for this second system which employs the valve panel of the previous machine, and a reversing valve 2 for the conveyors of both sections. The throttles 3 and 4 direct the oil simultaneously to the working conveyor mechanisms. If only one conveyor can be operated, in case of resetting or adjustment, the throttle of the second conveyor is set in the zero position.

The control of oil by simple throttles is recommended only in those cases where the forces to be overcome in moving the units are the same in both directions. If, for example, it is required to move two conveyors simultaneously in different directions, one loaded and one empty, it is advisable to use speed regulators with reducing valves; otherwise the speed will not be the same in both directions. The control device is shown as a rectangle 5, and the valve panel for inspection as a rectangle 6.

The electro-magnet of the unloading valve 7 is engaged and, after a short period, switched off so that, during a short time in each cycle, the fast-traverse pump 8 delivers oil into the system with partial overflow from the safety valve set at 355 lbf/in.^{2} (25 kgf/cm^2). Most of the time the oil goes back to the tank at the system pressure of 70 lbf/in.^{2}. (5 kgf/cm^2). This pressure is necessary to ensure normal working of the control device 5 and is the minimum for clamping the pallet on the machine. The speed of clamping and movement of the carrier bar mechanism is limited by the diaphragm-type hydraulic resistors 9 and 10. These consist of plugs with small holes, fitted directly into connecting pipes, and are applied in cases where there is no need for subsequent regulation or change of speed.

The conveyors X and Y of both sections are braked in both directions by the throttles 11 and 12 because both are working movements. Each drive controls two conveyors, longitudinal and transverse, coupled together by rack gear transmission A and B (Figure 7.4). When the loaded longitudinal conveyor moves forwards, the transverse conveyor returns empty,

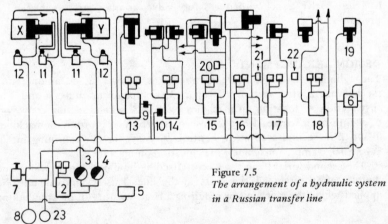

Figure 7.5
The arrangement of a hydraulic system in a Russian transfer line

and vice versa. The lay-out of Figure 7.5 includes reversing valves as follows: 13, for clamp adjusting; 14, for carrier bar movement; 15, for location on machines 1 and 2; 16, for clamping on these machines; 17, for location on machine 3; 18, for clamping on this machine; 19, for cylinder of the inspection device; 20, for the clamping mechanism for machines 1 and 2; 21, the pressure relay of the release machanism; 22, the pressure relay for clamping on machine 3; 23, the pump for the clamp mechanism.

INSPECTION AND FEELER DEVICES

Inspection and feeler devices are employed for checking dimensions or tolerances and the presence and depth of tapping holes. When the holes are sufficiently deep or the necessary dimensions have been achieved, the hydraulic cylinder moves the feelers forwards. These act, or cease to act, on terminal switches and allow the next machining operation to be carried out.

Figure 7.6 shows the unit for checking the depth of tapping holes. The inspection operation can coincide in many cases with the clamping operation, when at the beginning oil is pumped to the clamping cylinders, and simultaneously along pipes 3 and 4 to the hydraulic valve panel of the inspection device. The reverse valve 5 moves into the upper position in the diagram and directs oil through pipe 6 into cylinder 2 to operate the measuring rods 7 through a plunger. The plunger speed is limited by the throttle 8 and the measuring force is controlled by the reduction valve 9.

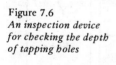

Figure 7.6
*An inspection device
for checking the depth
of tapping holes*

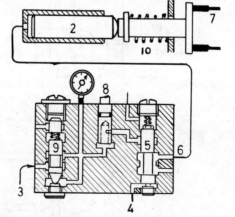

The plunger is returned to its starting position by a spring 10 but, if a differential cylinder is used, the plunger is returned simultaneously with the release of the clamp by oil fed directly from the pump. The piston speed of the inspection cylinder is sufficiently high at about 2 ft/min (0.6 m/min) because the operation can be carried out at any time during the cycle from clamping to release of component.

The pressure in the system beyond the reduction valve is regulated within the range from 100 to 200 lbf/in.^{2} (from 7 to 14 kgf/cm^2). A shaking or vibrating mechanism is sometimes fitted for removal of cuttings from drilled holes which require to be tapped. The system comprises a piston giving rapid knocks against the component, these being controlled by the amount of oil delivered and a throttle valve allowing from 0.5 to 8 c/s. One double traverse of the piston requires 2 in.^{3} (0.12 cm^3) of oil at a pressure of 300 lbf/in.^{2} (21 kgf/cm^2).

ELECTRONIC DETECTION OF DRILL BREAKAGES

One problem associated with transfer machines is that of detecting breakage of drills, particularly where drills break at the shank and are still held in position by the drill bushes. The problem is then to detect whether the drill is rotating or not. Automotive Products Group Ltd use a system which makes use of the fundamental principle of electricity whereby an electric current is produced when a magnet is rotated in close proximity with a coil of wire. A modification is used where the magnet is replaced by rotating the drill in proximity with a fixed magnet and a coil of wire, causing an electric current to be produced. This is fed into a fully transistorized unit incorporating an a.c. amplifier, and a full-wave rectifier for amplifying and rectifying the current for the operation of a relay (Figure 7.7).

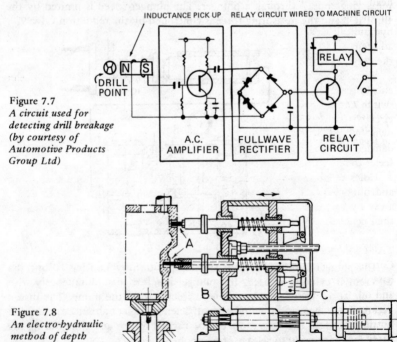

Figure 7.7
A circuit used for detecting drill breakage (by courtesy of Automotive Products Group Ltd)

Figure 7.8
An electro-hydraulic method of depth detecting

In order to differentiate between a sound drill and one that is broken, the signal is so arranged that the drill is only 'sensed' when it is being withdrawn from the component after the operation. By this means a 'fail-safe' condition is created, for, so long as an unbroken drill is rotating in proximity with the sensing device, a signal will be produced which allows the machine to continue working. A broken drill, however, will not produce a signal, in which case the relays operate and stop the machine. The sensitivity is such that the breakage of the flutes of a drill tip over about $\frac{1}{8}$ in. of the land can be sensed with reasonable certainty, even when submerged in coolant and swarf.

DEPTH CONTROL EQUIPMENT

A transfer machine which includes tapping usually requires a station for drilling, one for chamfering holes for easy starting of taps and a third for threading. The threads are generally produced by lead-screw-controlled spindles. When the full depth of thread has been reached, the direction of spindle rotation is reversed. If the hole has not been drilled to the correct depth, either the tap is broken or the spindle is damaged. Thus measurement of the depth of blind holes is important and necessitates the installation on the machine of equipment capable of checking every hole.

Figure 7.8 shows such equipment; the feelers A used for testing the depth are grouped according to the hole pattern in a measuring head. The head moves on ball guides B and forms the measuring unit which is advanced to the workpiece. Movement of the head is actuated electro-hydraulically, the impulse being obtained from a micro-switch on an adjacent clamping mechanism in order that the unit is brought forwards at the correct point in the automatic sequence of the machine.

When a hole is not the correct depth, the corresponding feeler encounters resistance and, as the head continues forwards, it operates a switch C through a rocking lever, thus bringing the machine to a stop. The circuit is arranged so that any feeler can bring the machine to a stop on its own, while an overrun device allows the stroke to continue without damage to the switch. The switch, however, operates immediately the feeler touches the bottom of the hole.

Holes to be tapped should not contain swarf from previous operations, and blind vertical holes are especially inconvenient. Swarf can be sucked away by vacuum equipment, and in some cases the feelers are bored down their centres and connected to a vacuum pump.

TAP-CLEANING EQUIPMENT

Cutting oil causes swarf to stick to taps and to clog them, so that devices have been introduced to spray each tap with a mixture of compressed air and oil. This is done while the tapping heads are in the retracted position. Figure 7.9 shows the principle on which air can be supplied from an air bottle or supply line. The supply line valve is opened by means of a cam D fitted to the base of the unit. Compressed air is led through small nozzles

Figure 7.9
*Tap-cleaning equip-
ment on a transfer
machine*

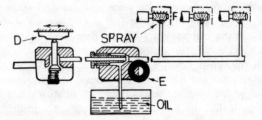

onto the taps which it cleans and lubricates simultaneously. The shape of
the spray which envelopes the tap is chosen to avoid the formation of oil
mist which might affect the operators. The mixing jet is shown at E with
guards F surrounding the spray jets.

DIAMETER CONTROL
On a transfer machine where many bores are machined, it is an advantage
to be able to measure the bores in order to check the quality of the work
and also to determine the point when it becomes necessary to change the
tool. The units are similar to those used for depth control but incorporate
Solex equipment with measuring heads fitted with air jets to be introduced
into the bores. For every hole two switches are required to trip at the
higher and lower limit of the tolerance range.

PRODUCTION TECHNIQUES ON TRANSFER LINES
Transfer lines are usually associated with many machining stations, but
pallet changing with only two loading and one machining station can be
used to increase work output. Figure 7.10 shows the DeVlieg jigmill

Figure 7.10
*A DeVlieg jigmill showing an automatic tool changer on the right
(by courtesy of Herbert Machine Tools Ltd)*

(Alfred Herbert Machine Tools Ltd) on which, using a normal single station, a component was completed in 20 min. By extending the bed on each side of the spindle, automatic pallet changing can be fitted so that work loading can be carried out on one station while machining is proceeding on the other one. This arrangement brings the production time down to 15 min.

As shown, the bedways for guiding the 360-position indexing work table on the X axis have been extended at the right and left for supporting two hydraulically operated slides for pallet changing. Each slide will take a pallet on which work is set and, for a pallet-changing sequence which follows completion of a machining cycle, both slides are advanced towards each other to be brought into contact with the table at opposite sides. At the end of this movement, plungers are depressed by the slides for releasing clamps which normally secure a pallet to the work table. At the same time, locating plungers are withdrawn, and the pallet on the work table and that on one of the slides are raised by an air flotation system.

The pallet carrying the completed workpiece is then transferred from the work table to the 'empty' slide by the pulling action of a built-in coupling unit. Simultaneously with this action, the pallet carrying the fresh part on the other slide is pushed onto the table. After locating plungers in the table have been moved upwards into engagement with bushes in the pallet, the slides are withdrawn. This action causes the clamps to be brought into operation for holding the pallet on the table, and it completes the pallet-changing sequence.

Pallet changing is thus carried out entirely automatically in a cycle time of 45s. The locating plungers enable a pallet to be positioned on the work table to an accuracy of 0.002 mm.

LINKED LINE FOR GEAR CUTTING MACHINES
Transfer lines are predominant for components requiring drilling, boring

Figure 7.11
A transfer line of Maxicut gear shaping machines
(by courtesy of Staveley Machine Tools Ltd)

Figure 7.12
An in-line transfer machine with 16 unit heads
(by courtesy of Fredk Pollard & Co. Ltd, Leicester)

and milling operations, but it is now realized that benefits can accrue with other types of machining operations; Figure 7.11 shows the transfer line for five Drummond Maxicut gear shaping machines (Staveley Machine Tools Ltd). These machines are in operation at the Longbridge Works of British Leyland Ltd, and they all cut a second-speed main shaft gear which has a 14 mm face width and 26 teeth at a helix angle of 30°. The pitch circle diameter is 152.4 mm, and the floor-to-floor time for the operation is 1.76 min.

Prior to passing down this line two Drummond Maxicut machines are employed to cut the clutch teeth on the same component. This has 30 teeth and a face width of 3.5 mm; the pitch circle diameter is 60.5 mm, and a cutting time of 20 s is achieved by the high stroking speed of 1420 strokes/min with a total floor-to-floor time of 29 s. As indicated, the blanks are conveyed along the front of the machines and are automatically loaded and unloaded at all seven stations. A swarf conveyor is also provided, leading to a central disposal point for filtration of the cutting oil.

Figure 7.13
A detailed view of the tooling

HIGH-PRODUCTION TRANSFER MACHINE

An in-line transfer machine designed and built by Frederick Pollard & Co. Ltd of Leicester is 30 ft long and has 16 spindle heads, some of which are set at compound angles relative to the track to within close tolerances. The unit heads are of various types comprising four air feed oil-checked units, four screw feed units, five air–oil drilling units, two larger air–oil drilling units and one air–oil tapping unit. These carry a total of 36 drilling and 25 tapping spindles.

A view of the machine from the first station towards the fixture at station six is shown in Figure 7.12, while Figure 7.13 shows a detailed view of five unit heads at station six during a machining cycle. The function of the machine is to perform multiple drilling and tapping operations on seven different basic designs of fuel pump bodies and to cater for 16 different variations as far as machining is concerned. In a number of instances, the drilling of small-diameter deep holes which must intersect accurately is required. Some of the drills used are so long and slender that it is necessary to engage them with their pilot bushes when they are stationary. Rotational drive to the drills is started only after they are safely within their bushes.

Each workpiece is mounted in a fixture which in turn is carried by a platen. This platen is supported on an air bearing while it is being moved down the track from station to station by means of a drag chain drive. As the platen arrives at a station, the air bearing is automatically cancelled, and wedge clamping by a pneumatically operated 'hammer blow' is applied.

There are five basic pre-set programme cycles for the machine, which can be selected by a rotary switch to suit the component being processed. Another feature of the machine is the use of a conditioning monitoring device. At the loading end of the machine there is a unit with nine push buttons, each labelled to indicate a particular cause of breakdown, or other reason for production being interrupted.

Within the unit there is a continuously driven paper disc and a recording pen. The disc makes one complete revolution per shift and, if the machine is idle for any reason, the operator pushes the appropriate button. The pen is thus shifted to the relevant position on the chart and continues to make a trace in that area until production is resumed and the operator returns it to the normal position.

One of the unit heads with a long slender drill can be seen at the bottom right-hand corner of the general view of the machines, while the air–hydraulic feed units are shown at the top of the near vertical larger heads on the left-hand side of the machine.

SAFETY ASPECTS AND MACHINE GUARDING

Photo-electric cell equipment has been developed to provide two screens around a machine, one around the extreme area and the other closer to the danger points. If the outer beam is broken, a hooter gives warning and simultaneously a danger notice is illuminated while the machine is stopped if the inner beam is broken. The operator cannot approach the tools without first switching the machine from 'auto' to 'inch' which puts the beam out of action. The same key which is used for this purpose has also to be turned to operate a switch at the station requiring attention before the inching button will operate and, as only one key is available, the operator is free from the danger that anyone else might start the machine.

Transfer machines do not readily lend themselves to conventional guarding; yet the necessity of protection for the operators is more pronounced than on standard machine tools for, in addition to the cutting tools, the advance and return movements of the heads does not take place simultaneously but at varying intervals of time. Similarly, units on a conveyor may suddenly 'shunt' without warning as one is removed from the delivery end and another is loaded.

Although a small number of operators is required to tend a transfer machine, these men are distributed along the length of the machine and may be out of sight of each other, so that distance plus noise of the machine prevents communication by speech. It is often that, under conditions such as these, the temptation arises for an operator to inspect the action of a cutting tool by moving between the heads, and he may be taken unawares by the sudden movement of mechanism and receive injury or worse.

BRITISH STANDARDS INSTITUTION CODE OF PRACTICE

From the illustrations in *Manufacturing Technology*: Basic and Advanced

Processes, it will be apparent that modern machine tools are well guarded as far as the mechanical parts are concerned. Nevertheless, the number of known accidents in factories per year averages about 272,000 with 550 fatalities. In view of this the BSI new code of practice *BS 5305, Safeguarding of Machinery*, has been published and should be studied by readers requiring a comprehensive survey of the subject. Another publication is *Health and Safety at Work*.

ROTARY TRANSFER MACHINES AND INDEXING TABLES

Rotary transfer machines posses the advantage of compactness when compared with an 'in-line' machine, and fixtures return to the starting point without the use of a conveyor, with practically no idle time in the cycle of operation. The limitation is the restriction in the number of work stations, usually about eight in the horizontal plane, but these can be supplemented by vertical or angular heads if the component requires them. Vertical rotary transfer machines usually have the unit heads mounted on a central column, but sometimes an external column is used. Other variations include a rotary indexing table or a fixed table with horizontal indexing heads.

Drum transfer machines are available, the construction comprising a number of fixtures bolted to and spaced around the periphery of a drum which is mounted on trunnions on a horizontal axis. The drum can be indexed and has one loading and unloading station so that one component is machined each time indexing takes place.

A typical rotary transfer machine with five unit heads is shown in Figure 7.14. These are spaced around the horizontal table so as to give a central loading and unloading station for the operator. There is a push button station and a pedal for clamping and unclamping the table which indexes under air flotation. The pedal is seen near the floor at the control station.

Figure 7.14
*A view of a rotary transfer
machine with five unit heads*

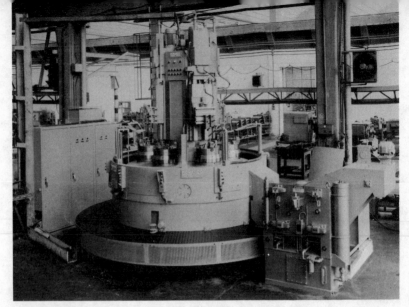

Figure 7.15
A rotary transfer chucking automatic machine. 260 hp is installed on the machine

The component is a simple cylindrical bar, 9 in. long, and clamping in the fixtures is by the air cylinders, horizontally and vertically mounted on the table. The operations are as follows:

Station 1: drill hole $\frac{3}{8}$ in. in diameter to 3 in. deep using incremental feed.

Station 2: drill hole, previously machined to 6 in. deep.

Station 3: complete hole to 9 in. deep.

Station 4: counterbore at one end.

Station 5: tap hole in counterbore.

Station 6: unload and load next component.

The incremental feed motion mentioned, sometimes known as 'woodpecker' feed, is necessary owing to the small diameter of the hole and length required.

It is essential that indexing tables are of the highest grade of workmanship and the design such that errors are not inherent in the mechanical features. One possible error is the radius of the locating plunger from the centre of the table, while another is the angular location between the index notches. A single error may not be of great significance, but cumulative errors may build up to affect the work accuracy.

Not all rotary transfer machines use indexing tables; for some operations such as milling or grinding can be performed on a continuous rotary table, but in general the size or type of work dictates the method used. However, considering indexing systems, the three most used are the Geneva motion, the cam mechanism or ratchet operation.

ROTARY TRANSFER TURNING MACHINES
Rotary transfer techniques are not restricted to drilling, milling and similar operations but are also extensively used for turning operations as well.

Figure 7.15 shows such a machine made by Flli Morando until the late 1960s. It has eight spindles, each mounted on a rotating drum which lifts to index and is located at each working positon by a large-diameter toothed coupling. Two stations are used for loading and unloading the component if double indexing is used which permits both sides of a fairly simple component to be machined on the same machine. This type of machine has fallen from favour to some degree as it is necessary to rough and finish one side completely before machining the second side. If a substantial amount of metal has to be removed from the second side of the part, distortion may occur and the geometry (roundness, etc.) of the first side may suffer in consequence. In-line transfer machines which are described in a later chapter elimate this possibility by turning the parts over three times, thus permitting roughing of each side before finishing of either takes place. Occasionally, turning over twice is sufficient to eliminate distortion.

The machine shown has separate drive motors for each spindle and separate feed motors for each head, thus permitting each machining station to be set up and tried independently of the others. Many contemporary machines had one motor which provided the drive for all the spindles and another which performed the same function for all the slide feeds; thus all elements of the machine had to be working during setting.

LOW-VOLUME WORK USING UNIT HEADS
Equally it should be stressed that unit heads may also be found in use in very-low-volume industries, their use not being confined to the kind of application found in the motor industry. An example is shown in Figure 7.16 where an 8 tonnes unit head is used to bore the main bearings of a 20,000 hp marine diesel engine. Boring bars up to 12 m in length can be used with a typical diameter of 350 mm. Up to 10 boring cutters are carried on the bar, one of which is seen on the operator's left; one of the engine main bearings and part of the engine bedplate can be seen on his right.

Figure 7.16
A large unit head operating on the bearings of a large diesel engine

Figure 7.17
A unit head mounted on a large lathe for boring operations

Typically, nine such bedplates are machined each year.

A further example of a unit head in use, on a standard machine in this case, is shown in Figure 7.17. The head has a mass of 8 tonnes and has a rotating spindle which can extend a total of 3 m. The spindle is hollow and carries a pilot bar which itself extends 3 m beyond the nose of the spindle; 50 hp is provided for the rotary drive, and 10 hp for the feed. The unit is mounted as a tailstock on the bed of a long large swing copying lathe; is turning the outside of marine diesel engine cylinder liners, the unit head engages its pilot bar with a bush in the lathe headstock spindle and proceeds to rough and to semi-finish the bore of the liner. Commonly the liners have a bore of 760 mm and a length of 2.5 m, the annual production being of the order of 200 parts. The bores are finally finished on a unit head machine which accommodates the liner in a vertical position so that its final machining is performed in the mode in which the liner will be mounted in the engine.

THE GENEVA (MALTESE CROSS) MECHANISM
Under ideal conditions a machine table should not move off the element which drives it at any point of its rotating path, and the Geneva motion fulfils this requirement. A good feature is that the speed of operation increases steadily from zero at the start of the movement and the speed is brought again to zero as the next station is approached. Thus one advantage is that the driven member cannot be overthrown as a consequence of its momentum.

A diagram of the motion is shown in Figure 7.18(a) which indicates a

Figure 7.18
*A diagram of the
Geneva mechanism*

(a)   (b)

roller entering a slot and leaving it (Figure 7.18(b)). A indicates the driving crank and B the locking plate. The torque of the driven shaft is transmitted to the driving shaft by the roller which bears against the leading side of the slot in the wheel. Since the roller is circular, the force applied to the wheel must act in a direction perpendicular to the centre line of the slot. At the instant of engagement, when the direction of motion of the roller centre passes the centre of the wheel, the line of action of roller thrust must pass through the axis of the driving shaft, ie tangential to the wheel. Hence at these instants the velocity ratio of the mechanism is zero.

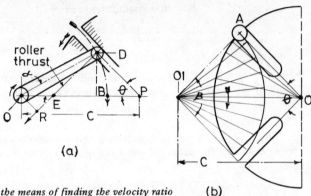

Figure 7.19
A diagram showing the means of finding the velocity ratio   (b)

To determine the velocity ratio at any intermediate position, from Figure 7.19(a) the velocity ratio is the ratio of the distances of driving and driven shaft axis from the line of action of the roller thrust; thus

$$\text{velocity ratio} = \frac{\text{angular velocity of wheel}}{\text{angular velocity of crank}} = \frac{PD}{OR}$$

However, the triangles OER and EPD are similar, since angle OER equals angle DEP and angle ORE equals angle PDE which is $90°$, therefore

$$\text{velocity ratio} = \frac{OE}{EP} \tag{7.1}$$

The maximum velocity ratio occurs when the line of action of the roller thrust is perpendicular to the common centre line OP of the driving and driven shafts. Thus it can be expressed as

$$\text{maximum velocity ratio} \; \frac{OB}{BP} = \frac{r}{C-r} \tag{7.2}$$

where r is the crank radius and C the centre distance of the shaft axis.

METHOD OF CONSTRUCTION
In laying out a Geneva mechanism it is necessary to know the diameter and number of slots. The procedure is then to draw a circle of radius OA

and to lay off the angle $\theta = 360/N$, where N is the number of slots. Bisect the angle θ by the line OO1 and through point A the line OA is drawn tangentially to the circle making an angle B/2 with the centre line OO1. The length of the driving arm AO1 and the centre distance C between the driver and follower are thus determined.

The maximum velocity ratio can also be determined analytically from Equation (7.2) by expressing the centre distance C in terms of the fixed angle between the two slots:

$$\text{maximum ratio} = \frac{r}{C - r}$$

$$C = r \cos A + \frac{r \sin A}{\tan B}$$

where A is the angle of crank from the common centre line and B the angle of the mating slot from the common centre line, both at the instant of roller engagement or disengagement. Also angle B = ½ × angle between two consecutive slots.

Since the centre line of the mating slot is tangential to the locus of the roller centre at the instant of roller engagement, then $A - B = 90°$. Thus

$$c = \frac{r}{\cos A} = \frac{r}{\sin B}$$

$$\text{maximum velocity ratio} = \frac{\cos A}{1 - \cos A} = \frac{\sin B}{1 - \sin B} \tag{7.3}$$

For example, the maximum velocity ratio for a five-mechanism is given by

$$\text{maximum velocity ratio (five slot)} = \frac{\sin 36°}{1 - \cos A} = \frac{0.5878}{0.4122} = 1.426$$

THE INVERSE GENEVA MECHANISM

In the inverse Geneva mechanism the directions of rotation of driving and driven shafts are the same, the configuration of a four-slot wheel is shown in Figure 7.20(a). The corners formed by the intersection of the slots must be cut away at least to a radius equal to the distance of the driving shaft axis from the roller. The action is kinematically identical with that of the ordinary Geneva mechanism, the roller engaging slots and driving through a fraction of a revolution according to the number of slots. No drive is transmitted while the roller is traversing the circular arc. The roller enters and leaves each slot tangentially, and hence the idling angle θ of the crank rotation during which no drive is transmitted is $(180 - 90°)$ or

$$\theta = 180 - \frac{360°}{n} = 180 \frac{n - 2}{n}° \tag{7.4}$$

where n is the number of slots. For example, the idling angle for an eight-slot wheel is

$$\theta_8 = 180\frac{8-2}{8} = 135°$$

LIMITATIONS OF ACCURACY

Although the Geneva motion is probably the best mechanism for accurate indexing, there are some limitations. Firstly, in regard to the circular locking profiles, if the load on the driven shaft is such that there is a torque acting on the slotted wheel during the locked period, there is a substantial bearing surface and with adequate lubrication this may cause no concern. However, at the point of the critical take-over to and from the driving pin, the condition is bad so that the locking surfaces can wear, resulting in the fact that the slots are not in proper alignment for engaging the driving pin. For this reason designers sometimes prefer to lock by means of a key or plunger actuated by an accentric or cam.

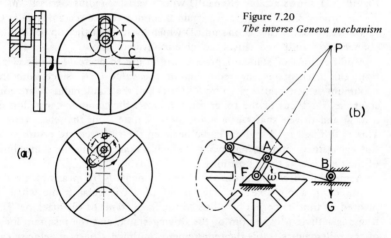

Figure 7.20
The inverse Geneva mechanism

(b)

(a)

There is, however, no reason why the path of a driving pin or roller need always be a circle. Figure 7.20(b) shows a four-station mechanism with a positive lock. The driving pin at D on the centre line of the coupler AB traces an oval-shaped path, symmetrical about FB. The input crank is now FA, and the output and input shafts are coaxial. The ratio of time locked to moving time is lower than that for the normal four-station external motion (2 to 1 for the case shown), and the maximum output angular speed is smaller. Accelerations are lower while locking is simple and positive.

The geometry depends upon finding the instant centre P of the coupler at the moment of engagement and ensuring that the line PD is then perpendicular to the slot in the output member (for all points of the coupler are, at this moment, moving perpendicularly to the lines joining them to P).

Three-, five- and six-station mechanisms can be designed on the same principle, but not two-station mechanisms.

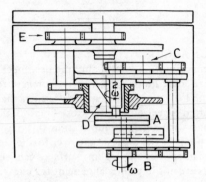

Figure 7.21
*The mechanism
of the Renault
indexing table*

PRACTICAL APPLICATIONS

Figure 7.21 shows a table (Renault) with a variable-radius Geneva arm at A; the driving shaft having a gear train B connected to a second train at C; this drives back to the second unit D which by teeth at the top of the hub connects to train E and causes the table to revolve.

Modifications of standard practice are in use to decrease the acceleration; this necessitates the reduction of the value $C-R$, since both the maximum velocity ratio given by $R/(C-R)$ and the crank radius R are constant, i.e. the locus of the roller must intersect the common centre line of driving and driven shafts at a point more remote from the wheel centre. This is fulfilled by a shallow curve between the roller centre positions at slot engagement and disengagement, which is tangential to the centre lines of the two slots at these positions.

A method of realizing the above requirements is by means of a roller mounted on a pin fixed at one corner of a triangular frame which is pivoted at another corner by a link to the fixed frame of the machine. The frame is actuated by the pin of the driving crank, and the resultant locus of the roller centre fulfils the requirements outlined. Constant velocity can also be obtained by introducing elliptical gearing into the design, the varying ratios between the different positions of the gears being used to modify the speed of the motion.

CAM SYSTEMS

Although the Geneva motions are effective for medium-speed applications, there are limitations.

(1) Its index angle is a fixed function of its number of stops. A four-stop must have $90°$ index angle, a six-stop $60°$, etc.

(2) Its acceleration diagram is fixed for a given number of stops.

(3) Its peak torque and its torque reversal occur when the control radius is nearly minimum.

Figure 7.22
*A diagram of a cam
indexing mechanism
(by courtesy of
Estuary Automation
Ltd, Shoeburyness)*

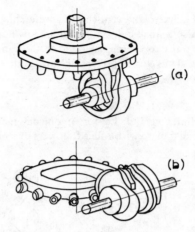

(4) It is not locked by pre-load during dwell.

The advantages of cam operation are that higher speeds are obtainable and with greater accuracy than with other systems. There are only two basic elements, the cam and the turret; Figure 7.22(a) and Figure 7.22(b) show two drives of the Automate indexing system (Estuary Automation Ltd, Shoeburyness). The form of the cam helix is a vital feature and a modified trapezoidal form is used to obtain a gradual change in acceleration at the beginning and end of the dial movement. The maximum force employed by this type of cam is proportional to its maximum acceleration calculated from maximum acceleration = 4.89 X stroke/index time. The constant 4.89 is the maximum coefficient of acceleration of the cam law. For rotary indexing, since torque = moment of inertia X acceleration, the inertial torque imposed on an indexing system with a known cam law is (1) proportional to the mass being indexed, (2) proportional to the square of the radius of gyration of that mass and (3) inversely proportional to the square of the index time.

Comparing Figure 7.22(a) with Figure 7.22(b), a convenient datum is the distance X between the turret and cam shaft centre. In the system shown in Figure 7.22(a) the pitch circle diameter of the cam follower will be equal to $2X$ but, in Figure 7.22(b) X must be shared between half the index cam diameter and half the pitch circle diameter of the cam followers. Thus the pitch circle diameter of the followers must be less than for those in Figure 7.22(a), about $1.3X$, so that, if a load W is applied by the cam track to a follower, the turret in Figure 7.22(a) will index with a torque some 50% greater than that of Figure 7.22(b). Thus the type in Figure 7.22(a) is suited for heavy-duty applications at speeds approaching 3600 operations/h, while in general that in Figure 7.22(b) is more suitable for higher speeds and lighter-duty projects.

With regard to design features shock forces can occur if the speed of the main cam shaft is allowed to vary by stored up energy in the dial plate due to acceleration during the first half of the index time. Thus a worm and wheel drive to the cam shaft is preferable, the ratio of the gears being 20 to 1. Also a small index angle with a low number of stops can cause the cross-over track to assume an angle to the cam shaft which would impair the action. Thus 40° is preferable for, the lower the angle, the smoother is the action.

ALTERNATIVE DESIGNS
Manifold Indexing Ltd, Leyton, in one design use a concave globoidal cam engaging a dial plate with roller followers extending radially from its edges,

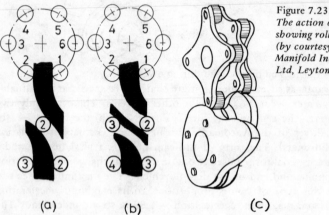

Figure 7.23
The action of indexing showing roller paths (by courtesy of Manifold Indexing Ltd, Leyton)

(a)   (b)   (c)

while an alternative is an arc-corrected barrel cam engaging a dial plate having axially disposed rollers. The shape of the cam is shown in Figure 7.23(a), in which the profile is unrolled showing six rollers with 1 and 2 in contact. As the cam rotates, before 1 loses contact, roller 2 is trapped to maintain control of the turret. Before 2 leaves the track, 3 contacts the cam and during the next dwell two rollers are again in contact.

If the cam is made as in Figure 7.23(b), two rollers pass through the cam during index. On the subsequent dwell, 3 replaces 1, and 4 replaces 2. This is a two-pass cam and the turret makes three stops in 1 rev; similarly an eight-roller two-pass mechanism is a four-stop. A parallel design is shown in Figure 7.23(c); this consists of a double cam with each track engaging a different set of rollers on a double turret. One set is displaced angularly from the others so that the followers alternately occupy equal angular divisions. The cam tracks are conjugate, and at any position of the cam shaft one track can exert a clockwise torque on a follower while the other operates anti-clockwise on another. The system is pre-loaded.

The next two chapters continue the subject but with more detailed descriptions of actual work handling.

8

ROBOTS AND OTHER
MATERIALS-HANDLING
SYSTEMS

Processing or machining in a factory almost always involves the removal of raw material from the inward point towards the point of despatch. Many machining operations and processes may take place in the course of this movement to change the raw material into the finished product, and during this process other items necessary to the running of the factory will also have to be moved around. Tools, fixtures, compounds and chemicals for heat treatment and finishing readily spring to mind. Inevitably, without mechanical aids a great deal of labour is involved in keeping these materials flowing and ensuring that they are present at the point where they are needed and at the time they are needed.

Frequent use is made in the mass production industries of mechanical handling of either parts or supplies, and it may be necessary in other industries for reasons other than the sheer volume of work passing through. The justification for mechanical handling can vary from a general need to reduce labour costs to a need for precision timing in the transfer of a part from one point to another. Equally, parts or materials can be dangerous to handle; red-hot parts from a furnace must be handled safely and there are many instances where parts must be inserted into an area where there is danger of a different sort. Perhaps a classic example of this is the need for complete remote handling of radio-active service parts for a nuclear reactor but, in general, labour reduction is the reason for the introduction of mechanized handling. Breakdowns excepted, a mechanism is always ready for work and performs at a pre-determined speed with consistency, and its running costs can be predicated with a fair degree of certainty.

From the examples given it will be seen that mechanical handling devices range from purely automatic units, either pre-programmed or computor controlled, through to devices which are remotely overseen or controlled by an operator from a remote point.

A full appraisal of the subject is beyond the scope of this chapter, as many of the devices in use are designed to handle one specific item, and there are certainly tens of thousands of items in this category. The chapter is concerned with the commonest types of mechanical handling devices of a wider or universal character and with a few examples of a more restricted nature. The first group to be examined is frequently known by the title 'robot', as it can perform many of the functions of the human arm and hand; it can perform these functions continually at a fixed rate and will undertake them in a hazardous environment. It often works faster and with greater precision than a man can, and it can be arranged to grip a wide range of workpieces (sometimes at random) and to load or unload a machine tool.

Figure 8.1
(a) *A Herbert mechanical arm linking two machines.* (b) *The arm arranged to mount a component between lathe centres (by courtesy of Alfred Herbert Machine Tools Ltd)*

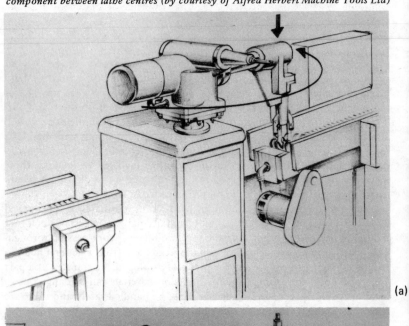

(a)

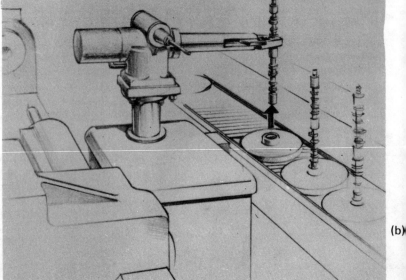

(b)

THE HERBERT MECHANICAL ARM

Operated electro-hydraulically by signals from a machine or from component-actuated micro-switches, this arm can be sequenced to link a pair of machines, to feed conveyors, to store parts or to transfer components in a wide variety of arrangements. Feedback signals from machines can be used to hold the robot or to stop another machine from producing further parts if the robot is already holding a part. The same signal can cause the robot to direct parts to a storage point or buffer. The fingers of the arm will open or close appropriately for either an inside or an outside grip; the wrist will twist through $180°$ and will also rise or fall along any swing position of the arm.

The unit consists of three free standing main assemblies

(1) The arm unit with its working height and swing restricted only by the weight of the components to be handled.

(2) An appropriate hydraulic power pack which may not be required if a suitable supply can be taken from other equipment nearby.

(3) A control console with hand controls for resetting purposes.

Figure 8.1 shows two applications. In figure 8.1(a) the arm is mounted on a cabinet between two conveyors, the arm lifting a component from an outlet conveyor on one machine to an inlet conveyor feeding the next machine, thus linking two machine tools. Figure 8.1(b) shows an example of a shaft-type workpiece which is lifted from a conveyor, is turned through $90°$ by the wrist action and is then moved by the swing of the arm to mount between the centres of the lathe seen on the left.

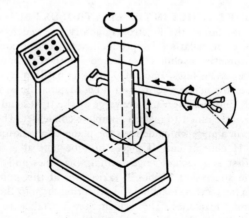

Figure 8.2
The Versatran industrial robot (by courtesy of Hawker Siddeley Dynamics Ltd)

THE VERSATRAN INDUSTRIAL ROBOT

The Versatran industrial robot (Hawker Siddeley Dynamics Ltd) (Figure 8.2) has been modularized into manipulator, controller, power pack and gripper, somewhat as the previous example. The manipulator usually consists of an arm mounted horizontally on a rotatable vertical post which in turn is mounted on a base. The arm can move horizontally and vertically, and it can be rotated horizontally through 240° by an electro-hydraulic servo system. The stroke in both lateral axes is 760 mm with an option of 1060 mm in the horizontal axis. The speeds are fixed at 910 mm/s and 90°/s swing. At the end of the arm is a wrist with two movements, sweep and rotate; servo controls facilitate positioning and this equipment is used for such diverse applications as welding and spray painting. Hydraulic rams, jacks and motors control the complete arm, and three solenoid-operated master valves can be closed to prevent any movement of the robot.

Punched disc programming can be used with this device, the disc being scanned by a photo-electric reader whilst a sophisticated manually set control permits up to 30 discrete arm positions to be pre-set in a sequence.

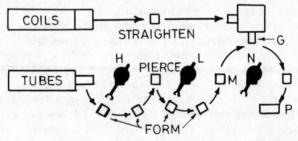

Figure 8.3
Three robots serving automatic machines in the Electrolux factory in Sweden

THE ROBOT LAY-OUT IN THE ELECTROLUX FACTORY IN SWEDEN

The robot lay-out in the Electrolux factory in Sweden is shown in Figure 8.3 and is an installation for producing refrigerator condensers in which several automatic machines are served by three robots. There are two parallel processing lines, one for producing baffle plates fed in coil form and another for making coils from tube. Material flows from left to right along each line and the parts converge at the right-hand end of each line where they are joined to make a complete condenser. The baffle plate line terminates in a press which delivers the plates to a pick-up point G.

Robot H collects cut-off pieces of tube, one at a time, and delivers them to first and second stage forming machines and to a hole-piercing machine in succession. Robot L takes over at this point, collecting the tubes from the piercing machine and delivering them to the third and fourth stage forming machines, from which they are taken to a buffer store at point M. Robot N then handles the finished formed coils and baffle plates which are assembled, joined and delivered to the finished assembly store P.

THE A.S.E.A. ELECTRICALLY DRIVEN ROBOT

One of the products of A.S.E.A. is used to illustrate a final example of the type of work which can be undertaken; the capacity of this device is much

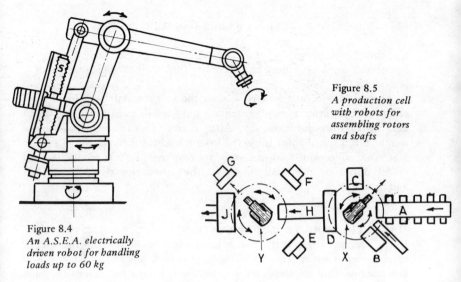

Figure 8.5
*A production cell
with robots for
assembling rotors
and shafts*

Figure 8.4
*An A.S.E.A. electrically
driven robot for handling
loads up to 60 kg*

greater than those previously described but should not be taken as an upper limit. Devices have been made to handle railway wheels during manufacture whose rough forged mass is approximately 1 tonne. Figure 8.4 shows this heavier robot which can handle masses of 60 kg. It has five movements and, despite being electrically powered, is quiet in operation and also is precise, its repeatability being within 0.4 mm. The control equipment for the robot is mini-computer based and permits movements to be programmed either on a point-to-point basis or as a continuously controlled path on three axes simultaneously, the speed of movement being infinitely variable.

The design comprises a jointed arm which is capable of rotary motion, mounted on a base, whilst movement in the elbow joint is achieved by two ball screws S, each coupled to a d.c. motor. One screw provides up and down movement whilst the other moves the arm in and out. Other movements are accomplished by means of d.c. motors and gear boxes.

A PRODUCTION CELL FOR ELECTRIC MOTOR ROTORS

A.S.E.A. have produced production cells for electric motor rotors for their own use, and the example, shown in Figure 8.5 shows robots helping to produce assembled and balanced rotors complete with their shafts. The key machine is the transfer line A which accepts previously sawn blanks which are transferred through the line by a walking beam to enable a series of turning and other operations to be performed. (The walking beam is described elsewhere in this chapter.)

Machined shafts from this line are fed into the area covered by robot X which has a complex programme as it must collect both shafts and rotor cores and must deliver them to the broaching machine B, the electric oven C and the assembly machine D. Each rotor on collection from the conveyor is turned into the vertical position before placing on the broaching machine. Each part is loaded and unloaded through the variouse processes by robot X until it is placed in the cooling tank H, from which the robot Y removes it.

The robot Y is programmed to serve the four remaining machines: E, F and G are cylindrical grinding machines and J is a balancing machine which also works fully automatically. After correction of any out-of-balance weight the finished rotor is ejected from the machine to roll down a chute ready for collection. Two men only are employed in this production cell whose duties are principally to ensure the correct operation of machines and robots.

MAGNETIC WORK TRANSFER SYSTEMS

During recent years there has been a rapid extension of the use of magnetic equipment for work handling; the high powers available to grip components can be illustrated by the use of magnetic chucks for such operations as turning and milling as well as the more common grinding. A careful analysis of the system requirements must be made in order to ensure a safe handling system, as the holding power of a magnet is affected by many variables.

(1) The workpiece material is very important as ferrites and soft iron may be held easily whilst austenitic stainless steels cannot be held at all.

(2) Good contact between magnet and workpiece is vital; quite small surface irregularities in the workpiece severely impair the magnet's hold.

(3) Both the thickness of the workpiece and the area of contact must be considered as well as its weight.

(4) The state of the material, whether heat treated or not, affects its ability to be held by a magnet.

The choice of a suitable magnetic holding device is one to be made with care, the maximum flux density which can be achieved in soft iron is about 20,000 gauss which gives a pull of 16 gf/cm^2. The gap between pole pieces of a magnet causes a reduction in the attainable flux density; an allowance of 15% is often made for this.

MAGNETIC ELEVATORS

For efficient continuous operation of a centreless grinding machine an automatic loading arrangement is essential. Figure 8.6 shows an automatic link between a centreless grinding machine and a duplex grinder to handle tappets (Figure 8.6(a)) (Wickman Scrivener Ltd) which are ground on the shank by the centreless machine and are surface ground on one end.

From the sequence diagram (Figure 8.6(b)) it can be seen that the parts loaded into the vibratory hopper A are passed (correctly aligned with the head foremost) to the push–feed arrangement of the centreless grinder, each part being pushed in turn on to the workrest. The wheel then advances for the automatic grinding cycle followed by its retraction for a

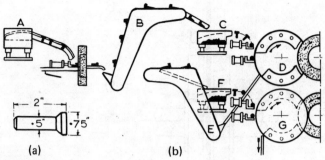

Figure 8.6
A magnetic elevator for feeding tappets to a centreless grinding machine
(by courtesy of Wickman Scrivener Ltd)

short distance so that the ground piece can be pushed out by the next piece arriving on the workrest. The ground piece lands on the magnetic elevator B which hold the part magnetically whilst transferring it to the next vibratory hopper C. The parts are again aligned correctly and are inserted into the side of a rotating carrier by an air-operated slide so that the tappet head contacts the side of the first wheel of the surface grinder D for rough machining.

After this operation the workpieces are fed into the hopper at the base of the second magnetic elevator E, and the process is repeated but using the second side of the rotating carrier and the second wheel of the machine. The equipment handles more than 450 parts/hr.

RUBBER BELT WORK TRANSFER

Most engineers are familiar with the use of rubber conveyor belts to carry diverse items from coal to precision engineered parts, this type of application having been in use for a great number of years. In more recent times rubber belts have been used in pairs to move items through various processes, the part being trapped between the two belts. A grinding operation is shown diagrammatically in Figure 8.7; the workpieces A are carried on a nylon or hard steel support bar B and are driven by the two belts which pass around the pulleys C. During the grinding operation the pieces are supported by the grinding wheels D whilst being pushed by the following component. The parts leave the machine through two guides located near the grinding wheels.

Circular, rectangular or odd-shaped parts can be handled in this way without damage provided that two reasonably flat surfaces, parallel to each other, are available to contact the belts. The belts are supported by the pressure plates E.

So far we have considered conveyors for one type of load only and with only one destination. Many distribution systems demand a wide variety of shapes and sizes of load and with the complication of several inputs to the system and several outlets from it. In the past mechanical guidance systems for each part have been used, but today parts can be routed

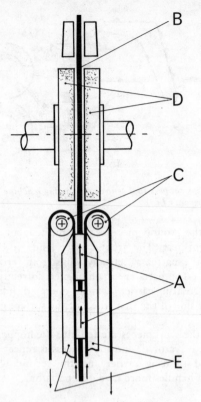

Figure 8.7
*The loading arrange-
ment of a twin disc
grinding machine*

through a system such as that described in a variety of ways. Often a small computer is used to close gates across certain parts of the conveyor system as an approaching part is recognized by sensors. A very effective means of doing this is described below which does not require a physical check to be made of the load at any point except at the input where each load is coded.

MAITROL SYSTEM FOR CONVEYOR NETWORK

The system of Brookhirst Igranic Ltd is one of a new non-contact memory, it is shown in diagram form in Figure 8.8. It provides a means of impressing a coded message on an object, so that, for example, a tote box can be automatically routed and conveyed to a pre-selected destination in a distribution network served by overhead, roller or belt conveyors. There are four basic elements in the system: a write station A, a code carrier B and a read station C as well as a read controller D. At each loading point, a write station serves to impress a magnetic code onto code carriers which are attached to the container that is to be moved by the conveyor.

The coded information is retained by the container during its passage through the distributing network from the initial write station to the intended destination point unless it is intentionally erased *en route* and is recorded in a different form at a subsequent write station. Each destination point in the conveyor network has an identifying code which is selected at the write station by means of a push button or selector switch. The magnetic write heads are immediately energized to produce the required code, and the information is then induced magnetically into the code carriers mounted on the side or bottom of the container as it moves past the write station. At the next and following read stations the magnetic address on the code carrier is monitored by reading elements and, when the code is recognized, a signal is transmitted to a gate actuator to divert the container along a conveyor branch track towards the destination point.

To return to handling devices which service machine tools the roller conveyors which feed parts to them must be considered. Many factories use simple conveyors with free rollers which do not require a full description. However, it should be stated that these suffer from several disadvantages: they must be inclined for the part to move without help, the machine tools themselves tend to govern the height at which each end of a conveyor can be placed and, if parts do not flow freely, manual intervention is necessary.

Accordingly, rollers which are driven by power are commonplace and generally work in such a way that, when a part arrives at the output, a sensor or micro-switch stops the conveyor to prevent the rollers from damaging the underneath surface of the component. As all the rollers usually have a common drive, this means that parts on it stop at regular intervals and often the system does not permit more than one part to be at the output at any one time; the constant stopping and starting of the conveyor under load can present maintainance problems. A powered roller conveyor can also be a safety hazard when heavy or bulky parts are being carried; tall parts can be unstable and fall when the conveyor starts and a positive drive to each roller can present a hazard if a person should become trapped by a roller or a part on the conveyor.

Another method has found many applications in recent years, the rollers being powered but not by a positive drive (Flli Morando and Co., Turin). A section through such a roller is shown in Figure 8.9. The shaft is driven continually by a chain sprocket B, one chain being common to all sprockets in a section of conveyor, whilst the drive is taken to the roller through a pair of bushes made from a friction material generally asbestos

Figure 8.8
The Maitrol system for a conveyor network
using a non-contact memory
(by courtesy of Brookhirst Igranic Ltd)

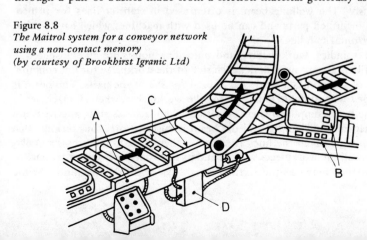

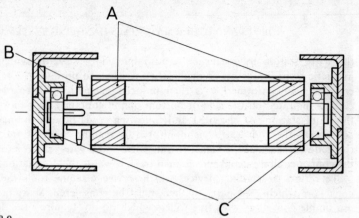

Figure 8.9
A section through a friction-driven roller from a conveyor
(by courtesy of Flli Morando & Co, Turin)

based A. All rollers are identical and when the conveyor delivers a part to a stop at the output, the rollers bearing that part will stop rotating as the bushes slip on the shaft whilst the remaining rollers continue to rotate and deliver more parts to the end of the conveyor. Thus all parts on a conveyor continue to move until they join a queue at the output whilst the drive motor never stops. The bearings C are also shown.

The length of bush in each roller is calculated to supply the necessary torque according to the workpieces to be handled; a very gentle handling of the load is assured by this Italian design as the conveyor speed can be slower because of its continuous running and the acceleration of each part is slow. The rollers can be made of soft metals or can be coated with plastics for delicate or highly finished parts. This latter form of conveyor is ideal for delivering parts to a transfer line where a precise pick-up point is needed by the line's own handling system, the friction-driven roller keeping parts against a stop without damaging them.

THE WALKING BEAM
The pick-up point of a conveyor is often linked to a walking beam transfer device which can undertake the precise movement of several parts simultaneously. The walking beam is often used in transfer lines for turned, milled or drilled parts and can be used with machines which have vertical or horizontal spindles.

The spindles, work-loading and unloading points are all equispaced; a pair of bars are arranged under each side of the workpiece for flat components and sometimes a single bar is used for shaft-type parts. The beam is equipped with suitable locations for the workpieces; when all machines in the line have completed their operations the machines stop, the parts are automatically unclamped and the beam is made to rise, thus picking each part up which is in the line. The beam is then made to move horizontally, carrying all the workpieces with it, and its horizontal movement is equal to the distance between the fixtures or chucks on all the machines. At the

Figure 8.10
A walking beam in a transfer line for cam shafts
(by courtesy of Industrial Sales Ltd, Wilmslow)

end of its stroke the beam descends and deposits each piece in the machine following the one that it has just left. The last piece from the line being deposited on an output conveyor and the first station on the walking beam has picked up a part from the input conveyor and has put it in the first machine.

The pieces are clamped and each machine starts its cycle again whilst the walking beam returns to its original position.

A walking beam is occasionally referred to as a 'lift and carry' system and Figure 8.10 illustrates this. The final five cam shafts in a turning transfer line can be seen, the most distant one being between the centres of the

Figure 8.11
The running centre is positioned to push one cam shaft into the chuck and to support the part after the chuck has gripped it. The supporting fork withdraws before machining

final special purpose lathe, whilst the remainder are held in fixed fingers on the main frame. Inside the main frame can be seen the moving fingers attached to the walking beam which is a single one in this case; it lifts the cam shafts, traverses them forwards to the next set of fixed fingers where it inserts them and returns to the start position. Figure 8.11 shows the beam passing through one of the lathes in the line.

In a very long transfer line the walking beam may be split up into several independent lengths with component storage in between each section. This allows a degree of flexibility in the line and permits tools to be changed on a particular machine or section without stopping the complete line; automatic inspection is often undertaken at these points.

A part of a view of a walking beam from a transfer line is shown in Figure 8.12, the machines in the line being twin spindle vertical chucking automatics followed by vertical drilling and vertical reaming machines. At the top of the illustration can be seen two 'nests' into which the rough castings (motor car brake discs) are loaded by robot and the bars which comprise the walking beam can be seen in front of and behind the nests. At the start of workpiece transfer the bell-crank shown in the bottom right-hand corner rotates clockwise through 75° and is coupled to another at the opposite end of the beam to ensure that the beam is lifted in a perfectly horizontal position. The locations for the workpieces on the beam are mounted on its upper surfaces and hold the part loosely with, perhaps, a clearance of 1.5 mm to 2.5 mm on the diameter. After lifting, the beam is moved horizontally as previously described; the movement is provided by a hydraulic cylinder whilst the beam runs on rollers located on top of the bell-cranks. As the machines are twin spindle types, two

Figure 8.12
*One of a pair of elevating units for a walking beam
(by courtesy of Flli Morando & Co, Turin)*

parts are moved to each machine at one stroke of the beam; reverse rotation of the bell-cranks serves to lower the parts into the chucks and fixtures whilst the beam continues to descend for a further short distance to clear the work. The beam then returns to its start position.

THE WALKING BEAM FOR A GRINDING MACHINE

An example of the walking beam applied to grinding operations is shown in Figure 8.13; the machine is a Cincinnati Milacron centreless grinding machine. A gantry-like structure spans the working area of the machine and has a horizontal beam A which is arranged to rise and fall, actuated by cylinders B. A square sectioned shaft C is secured to the underside of the beam and is arranged to reciprocate horizontally relative to the beam. Reciprocation is effected by the hydraulic cylinder D and the extent of travel is indicated by the solid and dotted lines.

The square sectioned shaft rises and falls with the beam A, the uppermost position being indicated by C and its lower position by C_1. Attached to the square shaft and projecting downwards from it are four pairs of hooked workpiece carriers E. The shaft C can also be swivelled about its

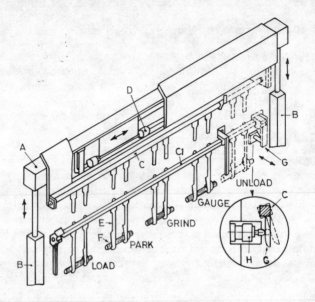

Figure 8.13
An example of a walking beam applied to grinding operations
(by courtesy of Cincinnati Milacron Ltd)

longitudinal axis through a limited angle. The cylinder H applies pressure
to lever G at the appropriate point in the cycle to swivel the shaft and to
disengage the hooks E from the workpieces. After the return of the beam
in a horizontal manner the hooks are allowed to swing to their original
positions to re-engage the next workpieces.

Thus by this sequential pattern of movements the workpieces are
advanced through the various positions shown. 200 parts/h are handled
and ground in the arrangement shown, the parts being distributor shafts
for motor cars and 334 mm long.

Figure 8.14
A gantry loader for a crankpin grinding machine
(by courtesy of Landis–Lund Ltd, Crossbills)

GANTRY LOADING SYSTEM

The automatic work loading in Figure 8.14 is by gantry loader of a Landis crankpin grinding machine, the workpiece being a tractor crankshaft. The gantry which spans the machine is located on top of two electrical cabinets which house solid-state control equipment for both machine and loader. A powered carriage runs on the gantry which has fingers able to extend and grip the component. The carriage carries components between the conveyor shown on the right and the machine. There are 13 positions on the conveyor which is a true lift and carry device, six to accommodate ground shafts and six to accommodate unground ones, the centre position being used for the carriage pick-up and drop-off station. The loader–carriage will remove a part from the machine; the conveyor will move forwards one position to take the ground part from the centre station and to bring an unground part forwards; the carriage then picks this part up and delivers it to the work position on the machine. In the example shown the loader is operated manually but many such devices are in automatic operation.

Gantry loaders are also used for similar applications on high-production lathes and sometimes the carriage is arranged to have a loading arm on both the front and the back. In this way two parts can be carried at once, the unmachined part and the finished one. The lift and carry conveyor would then have two positions in its centre, one for the part to be loaded and the other for the part being unloaded. Where high-volume manufacture is undertaken such a device will permit the load–unload cycle time to be nearly halved.

AN INVERTING DEVICE

Large flat components machined on transfer lines may need to be turned over several times in their progress down the line, such parts being of the brake drum, brake disc or flywheel types. A number of such devices are in use and Figure 8.15 shows a particularly good example (Flli Morando and Co., Turin). Devices are available to load chucks, as has already been shown, but this example unloads from one chuck, inverts the part and also loads it into a second chuck (or onto a conveyor). Three functions are therefore performed by one unit, establishing a saving on cost, complexity and floor-to-floor time.

The transfer line consists of two twin-spindle Morando vertical chucking automatics; the first one roughs each side of a brake disc and the second one finishes each side of the same part, intermachine handling being by gravity conveyor.

The turn-over unit is shown diagramatically in Figure 8.16. After the chucks have stopped at the end of a cycle in a pre-determined radial position, the fork is advanced hydraulically in the direction A to embrace the workpiece and then the chuck jaws open automatically. After this, the inverting device tilts on a bar as shown by the arrow B and this action serves to lift the part from the chuck. Normally this inclination is only

Figure 8.15
*A mini-transfer line for machining brake discs
(by courtesy of Flli Morando & Co, Turin)*

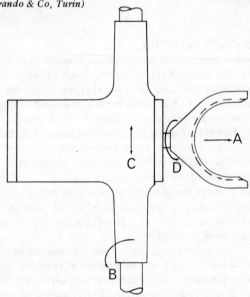

Figure 8.16
A workpiece-inverting device: A, collect workpiece; B, Lift; C, transfer; D, invert

sufficient to clear the chuck and to allow gravity to contain the workpiece within the fork, but in some instances the fork is tilted until it is vertical. A hydraulic ram then moves the whole device with its workpiece in the direction C whilst a rack within the device rotates the fork as shown at D. The rotation is arranged to be 180° during a travel equal to the distance between the two chucks. The part is thus inverted and positioned over the second chuck; the tilt is then reversed to lower the part into the chuck which closes and the fork withdraws. Turning commences whilst the inverting device returns to its original position.

Occasionally, shaft-type parts require turning end for end during transfer line machining; this type of device finds applications in that area also.

WORK STORAGE (ON LINE)

Mention has already been made of the benefits which accrue from buffer storage installed at key points in the line. The physical appearance of a store is usually dependent on the nature and size of the part being processed but local conditions may dictate the actual configuration of the storage area. Perhaps the simplest example might be of a long length of powered conveyor between two machines in a line; the conveyor can be 'folded' to make it conform to the area available for it and such an arrangement is frequently used for flat components as in the previous example. Factory floor space is usually expensive and in recent years installations have appeared which store the parts in a vertical manner, either by turning flat parts through 90° and passing them through the store on their edges or by folding the conveyor in a vertical fashion, usually in a spiral.

TOWER STORAGE

The parts to be drilled and bored in the special purpose machine (Unistand, Bologna) shown (Figure 8.17) would be delivered by a powered conveyor system (not shown) to the top of the spiral runway in the storage system. Gravity is then often used to feed the parts to the machine down the runway but occasionally the runway is made to vibrate to assist the movement of the parts. In some cases an escapement mechanism is incorporated to ensure that the parts are released singly. In the example shown in Figure 8.17 vibration is affected pneumatically.

PALLETIZING

The work handling described so far in this chapter has been of a type where the part is itself rotated during its principal machining and is of a robust nature, thus allowing it to be handled directly. There are many machining transfer line applications where the part is perhaps of a cubic shape or where several parts have to be held together in a group whilst machining is performed. Certainly, also, there are many parts whose shape

Figure 8.17
A tower storage system (by courtesy of Unistand, Bologna)

does not allow them to be moved by conveyor or whose basic material is aluminium which might not allow handling on a conveyor because of the possibility of damage. Those parts of a cubic type which require milling, drilling and tapping also present problems of location as tolerances may demand all machining to be done at one setting whilst production requirements may preclude all but a transfer line approach.

Work of these types is often clamped on a fixture or pallet, sometimes after a pre-machining operation to provide a location. The fixture or pallet then travels through all the machining operations with the part, and each machine in the line uses locations on the pallet prior to clamping rather than on the part. The work is thus protected against damage during transit, and accuracy of machining can be ensured.

The Kamaz truck factory in Russia employs such a system for its rear axle machining; the axle housings are clamped in their rough state onto a pallet before machining and are removed afterwards, the pallet being returned to the start of the line by overhead conveyor with a station where the pallet is inverted, pressure washed and dried before its arrival back at the start of the line; all cuttings are removed in the process.

CAR BODY TRANSFER LINE

An innovation in the manufacture of car bodies has been introduced by

Comau Industriale of Turin; it uses new principles of handling and welding combining robots for welding and palletized handling and eliminates many of the problems associated with this product. Previously most car bodies were made on 'rigid' transfer lines which could only accommodate one type of body with minor variations to it such as the inclusion of a sunshine roof or a four-door version instead of a two-door type.

This entirely new approach to the car body is shown in Figure 8.18,

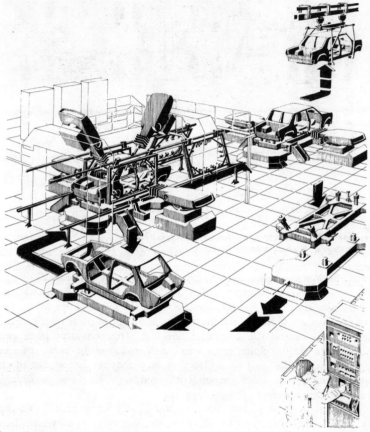

Figure 8.18
An impression of robot trolleys and robot welding of car bodies
(by courtesy of Comau Industriale, Turin)

and in addition to the robot welders the pallets are carried on robot trolleys which are guided from beneath the floor by high-frequency signals. Each trolley is able to identify itself to the on-line computer which controls all aspects of production in the body shop. There are many

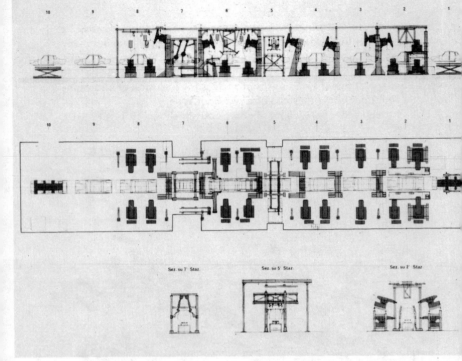

Figure 8.19
*A section and a plan view of the Volvo car welding line
(by courtesy of Comau Industriale, Turin)*

flexible welding stations, but any one body may only visit a small number of them and several different bodies can be manufactured simultaneously, each visiting only that station appropriate to the required operation. Two such stations are shown in the illustration and a robot trolley with a pallet can be seen behind the computer terminal. The computer plans each operation for each welding station and each trolley using stored information; its overall planning of production only requires knowledge of the type of body panels which are being delivered to the shop and the order in which they are arriving.

The result is a completely flexible body factory whose plant can accept any of the company's types including future designs; previously a rigid line had to be scrapped when the model that it manufactured was discontinued.

Figure 8.19 shows a mixture of the original and the new method of manufacture, a rigidly laid down line but one which employs eighteen robots to carry the spot welders; it is able to handle all the current models of Volvo made at the Swedish plant and was also supplied by Comau Industriale. As in the previous example, only fixturing and programming need be changed to accept a new model of car body. 60 units/h are produced.

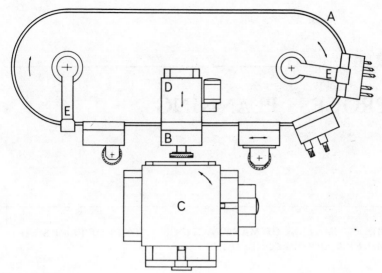

Figure 8.20
A diagram showing the travelling head machine for the Russian tractor chassis

As has been mentioned already earlier in this chapter, handling of jigs, fixtures and tools can also present sizable problems in a factory. Where the workpiece and machining are relatively light the problem may not be too severe, but where large items are to be handled there are cogent reasons to avoid moving parts around the factory which have a mass of several tons.

TRAVELLING HEAD TRANSFER SYSTEM

A logical approach to this problem is to arrange for the machines to move to the workpiece so that the latter remains stationary, and this is shown diagrammatically in Figure 8.20. The diagram shows a track A installed well above floor level which carries a number of unit heads without motor drive B. These are able to travel around the track, assisted round the bends by the two arms E, and to be positioned in front of the motorized drive unit D which automatically connects itself. The workpiece is situated on the table C which is provided with sliding and rotary movements. Thus, once the workpiece is fitted in position, a large number of heads may be brought to work on it, sequentially, and all machining may be completed at one setting.

A complex such as this is to be installed in the USSR at the Ceboksary plant for making the main chassis of crawler tractors. The workpiece is some 18 feet in length and approximately 25 heads will be contained by the track. Because the complete group is numerically controlled, one operator only will be required; a mechanical loader is also being installed to put the chassis onto the table.

Tool changers on smaller machines could also be considered within the framework of this chapter but are described elsewhere in this book, together with their respective machines.

9

PROCESS PLANNING

THE ESTIMATION OF MANUFACTURING COSTS OF PUMP SHAFTS UNDER VARYING CONDITIONS

It is normally expected that the engineering industry functions for commercial reasons and therefore there should be constant effort to reduce manufacturing costs in order to maintain or improve the company's position relative to its competitors. Equally, a new product may be manufactured in pre-production quantities on existing machinery but new plant is frequently introduced to ensure manufacture at the lowest cost commensurate with the desired quality of product.

The choice of suitable machine tools for a new project is often a difficult one to make; a wide range of different manufacturing methods may be open to the user with equipment available from many manufacturers for each method. There may well be many different ways in which a part can be held during machining and, if there are several operations, it may be possible to arrange the sequence in various ways. These facts, together with the staggering array of cutting tools available, serve to add confusion to an already complex situation.

The machining method may, for a new product, be based on previous experience with a similar product but its basis is usually formed around predicted sales. It is necessary to examine the effects of faulty sales forecasts in order to realize their importance. To take a simple case, we may consider the situation of a company about to manufacture a new product, say a pump, and the forecasted sales for the unit are approximately 10,000 units/year. Examination of the shaft may show that it can be turned on a centre lathe in 30 min and, if the lathe is employed on 15 shifts/week (7½ h shifts), 10,000 parts can be made in a year. If the machine is bought and sales reach expectations all should be well, as the planned machining costs will be achieved, perhaps in the manner set out below.

Company policy demands that the initial cost of the machine and any capital charges made should be recovered over a period of 3 years so, if these total costs are £30,000, an annual fixed charge of £10,000 must be spread over the units made.

Therefore

machine and capital charges (£10,000 per annum)	= £1.00 per part
operators' wages and factory overheads (£6.00 per hour)	= £3.00
total cost per part	= £4.00

Let us now assume that sales only reach 3000 units/year; then we have machine and capital charges remain the same

(£10,000 per annum)	= £3.33 per part
operators' wages and factory overheads (£6.00 per hour)	= £3.00
total cost per part	= £6.33

The result of a sales overestimate in the case quoted increases the price of one part of the pump by nearly 60%, which would probably be the situation with other parts of the pump too. Such a result would probably overprice the product and the company would therefore have great difficulty in selling it. Different companies have different methods of allocating a figure to the cost for running a machine but it will be clear, in the simple case quoted, that the expected performance is far from being achieved.

In the case of an underestimate the position can be just as serious, delivery promises for the complete unit may not be met and as sales rise, say, first to 20,000 units/year and then to 30,000 there may be the temptation to buy first one extra and then a third centre lathe to cope with the large order influx. In that case the machining cost would remain as in the first example quoted above, £4.00 per part if the machine cost had not increased between buying the first and subsequent machines. This is still an unsatisfactory state of affairs, however, as use of a more efficient manufacturing method would reduce costs, a desirable situation permitting the possibility of a somewhat stronger marketing position.

The installation of, say, an automatic multi-tool lathe or a copy turning lathe at the initiation of the project (with a capability of 30,000 parts/year) would have resulted in expensive shafts in the early stages but in much cheaper shafts when sales had reached 30,000 per year. Such machines might make the shaft in 10 min, ie 6 per hour, and the previous calculations would appear as follows:

first-year sales	10,000 units
second-year sales	20,000 units
third-year sales	30,000 units
Machine cost including capital charges	£35,000 per year

The machine can run a single shift during its first year, two shifts during the second and three shifts during its third year.

During the first year

machine and capital charges	$\dfrac{35,000}{10,000}$	= £3.50 per part
operators' wages and factory overheads	$\dfrac{\pounds6 \text{ per hour}}{6 \text{ parts per hr}}$	= 1.00 per part
	total cost per part	= £4.50

During the second year

machine and capital charges	$\dfrac{35,000}{20,000}$	= £1.75 per part
operators' wages and factory overheads		= 1.00 per part
	total cost per part	= £2.75

During the third year

machine and capital charges	$\dfrac{35,000}{30,000}$	= £1.17 per part
operators' wages and factory overheads		= 1.00 per part
	total cost per part	= £2.17

Quite apart from the obvious financial savings other benefits should accrue from this automatic lathe approach to the shaft, only three operators being needed (one per shift) plus part of a setter's time instead of nine operators for the centre lathe approach. In the latter case unskilled operators could probably be used at a slightly lower cost where it would not be usual to do so for centre lathes.

It should be noted that the shafts made in the first year cost 0.50p more than they would have done when made on a centre lathe but in the following two years they are £1.25 and nearly £2.00 cheaper respectively.

It would perhaps be fitting to add that a further substantial increase in requirements for these shafts would permit an even more sophisticated approach to their manufacture; a multi-spindle automatic lathe or even a transfer line could be considered. A multi-spindle machine capable of making 200,000 shafts/year could incur annual machine plus capital charges of £234,000 without this exceeding a charge of £1.17 per part as in the previous example (third year). It might be that the hypothetical pump that we have considered in this example has a total of ten or a dozen major parts requiring machining; the possible consequences of incorrect sales forecasts can only be substantial losses as has already been shown; often a product must be manufactured at or under a target price and the possible variations shown in the example could not be tolerated in a commercial environment.

It will also be seen from the example that many of the possible machining methods available to the production engineer are frequently unusable as the sales forecast effectively restricts his choice of machine.

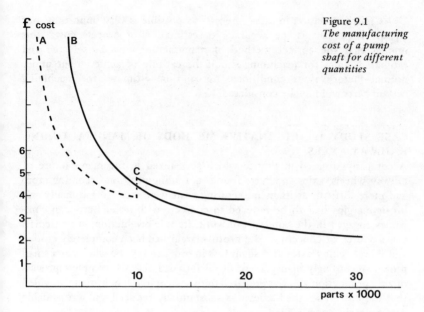

£ cost

Figure 9.1
*The manufacturing
cost of a pump
shaft for different
quantities*

Figure 9.1 illustrates the example by plotting the cost per component against annual output. Curve A is the centre lathe and curve B represents the automatic lathe. A further curve could be added for a multi-spindle automatic lathe, permitting comparison of the three possible methods considered; at all times the assumption has been made that the machines could not be occupied on other work if there was a shortfall of orders but that their operators could be occupied elsewhere.

A final note on the consequences of overestimates should point out the danger point at 10,000 parts/year on the graph of Figure 9.1. A reasonable degree of confidence in a forecast of 10,000 parts/year to manufacture might tempt a production engineer to recommend the acquisition of machine B in the belief that subsequent years would produce production quantities well in excess of this figure and thus justify the purchase of machine B. The graph shows that annual production would have to rise to 13,000 per year to show a saving on each shaft of 0.31p whilst a shortfall of the same magnitude (3000 parts) would result in each shaft costing £2.00 more to turn than the predicted centre lathe cost of £4.00.

On the other hand, if machine A was purchased, the same shortfall would only increase the machining cost of each shaft by 0.42p whilst an annual requirement of 13,000 shafts would require a second centre lathe to be bought C; the shafts made by the second machine (3,000 in a year) would cost £6.33 each. The total production (10,000 parts at £4.00 each and 3,000 parts at £6.33 each) average cost would be £4.54 per shaft. Therefore, as soon as production requirements call for a second centre lathe, machine B becomes economic.

As much precision in sales forecasts as possible is very important, but equally important is the accuracy of estimation of process costs; this implies that the correct method of manufacture must be selected and estimated times for machining should be capable of achievement under normal factory floor conditions. Actual time estimates for machining certain parts will now be considered.

CASE STUDY 1: ALTERNATIVE METHODS OF MANUFACTURING RAILWAY AXLES

A company situated in France which specialized in the manufacture of railway wheels, axles and tyres found its trading position becoming more and more difficult as its manufacturing costs were escalating annually and profit margins had to be reduced to compete with other European companies engaged in the same sort of work. Its axle production, in particular, was a source of concern as the profit margin had been completely eroded and it was selling axles at a slight loss in order to sell the wheels at a small profit. Accordingly management decided to examine the complete process of axle production and was prepared to invest capital in new plant provided that the cost of finished axles was substantially reduced and a reasonable profit margin was provided.

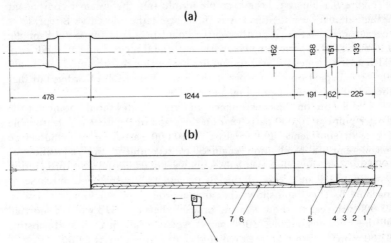

Figure 9.2
(a) *The finished dimensions of a railway axle.* (b) *The sequence of cuts for a single-slide copying lathe, third operation*

Axles were made in various quantities from 'one-off' items which were for repairs to existing vehicles to production runs of up to 5,000 of a type in one order. Market research indicated that the small quantities could be sold competitively at a higher price than had previously been charged as customers were prepared to pay a premium for these 'one-off' items.

Attention was therefore concentrated on the larger batches which were frequently ordered and each operation was examined to find where savings could be made. The sequence of operations was as follows (after forging).

(1) Mill ends to length.
(2) Centre drill.
(3) Rough turn first end.
(4) Rough turn second end.
(5) Finish turn.
(6) Re-cut centre holes.
(7) Grind bearing and wheel seatings.

Examination of these operations showed that the greatest savings could be shown on the longest operations, (3), (4), (5) and (7).

The rough turning operation was being performed on a variety of lathes, some with copy turning facilities and some without; the timing of similar axles over the years had shown that the copy lathes were much faster than the conventional machines which had to produce the radii by form tooling methods and were laboriously slow. It was decided to consider the replacement of the latter category of machines with a more modern type which embodied copy turning and as much power as could be usefully used on an axle.

An examination of acceptable copy lathes with a swing of around 300 mm showed that few had power in excess of 35 hp available from the main motor and in most cases the power available at the chuck was between 20 and 25 hp. The existing machines had 20 hp motors. (See *Manufacturing Technology*, Volume 1, *Basic Machines and Processes*, Chapter 11, for typical losses in a lathe headstock.) Several axles which had been machined in large batches previously were studied and machining lay-outs were prepared to produce estimated times for their machining on the proposed new machines. The higher power available permitted higher rates of feed to be used and, for some cuts, deeper cuts could be used with the same feed, rate thus reducing the number of cuts needed. Figure 9.2(a) shows one of the axles (finished dimensions) which had to be machined from parallel forged bar and Figure 9.2(b) shows the machining method adopted for the first end of the axle. (The forged bar is 200 mm in diameter.)

The forged bar was held between the lathe centres whilst it was driven by a clamped-on driver at the headstock end. Parallel cuts were used (without the copying facility) for cuts 1, 2, 3 and 6 whilst the copy turning was used for the final roughing cuts 4, 5 and 7. The machining times with speeds and feeds are given in Table 9.1. It will be noted that at this stage (third operation) the axle is more than half roughed out and only the 188 mm, 151 mm and 133 mm diameters need to be turned on the fourth operation with a blending cut on the 162 mm diameter where the tool had been led into the work at an angle of $8°$ to permit proper tool clearance. The estimated times for the fourth operation are given in Table 9.2.

The total estimated machining time for the third and fourth operations, $76.95 + 28.9 = 105.85$ min, was only 21% better than times currently

Table 9.1
Cutting data, first end of railway axle, rough turning (single-slide copying lathe)

Operation	Turned diameter (mm)	Rotational speed (rev/min)	Cutting speed (m/min)	Feed (mm/rev)	Feed (mm/min)	Radial depth of cut (mm)	Saddle travel (mm)	Cutting time (min)	Idle time (min)	Total time (min)
Cut 1	183	125	72	0.4	50	8.5	285	5.70	0.60	6.30
Cut 2	166	125	65	0.4	50	8.5	255	5.10	0.60	5.70
Cut 3	149	157	74	0.4	62.8	8.5	222	3.50	0.60	4.10
Cut 4	133	157	93 (maximum)	0.4	62.8	8.0	287	4.60	0.85	5.45
Cut 5	188	125	74	0.4	50	60	192	3.85	0.40	4.25
Cut 6	181	125	71	0.4	50	95	1,244	24.90	0.60	25.50
Cut 7	162	125	64	0.4	50	95	1,244	24.90	0.75	25.65
Total for all cuts, third operation										76.95

achieved on the older copy lathes and would not justify the cost of new machines with their higher hourly rate. That is there would not be a saving at all in real cost of machining the axle on the third and fourth operations. Other axles which were studied showed very similar results.

It was clear that there would be an improvement in machining times and that the improvement was attributable to the extra power available; even shorter machining times were required for little extra cost if new machines were to be purchased and the company were to stay in business. An extensive survey of copying lathes was then conducted which revealed that only one machine was known which had a substantially higher power, 40 hp. The machine cost 30% more than the one concerned with the time study in Figure 9.2 and, as machining times could not be expected to be reduced by more than a further 10%, enquiries were made in another direction. Could the latest cutting tools promise a further reduction in time? Could larger tips on the tools permit deeper cuts? The answer again was a requirement for increased power.

Further reference to Figure 9.2(b) shows that, for example, a second tool mounted behind the one shown and 9.5 mm closer to the work would permit most of cuts 6 and 7 to be made simultaneously but again extra power was required. A machine tool builder was asked to consider the manufacture of a machine specially for axles and with the ability to use several tools at once. The result was the machine shown in Figure 9.3.

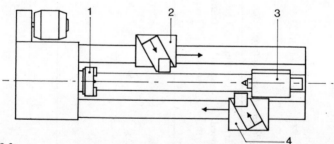

Figure 9.3
A special automatic twin-slide copying lathe for railway axles: 1, chuck with twin floating jaws and fixed centre; 2, rear copying slide; 3, hydraulic tailstock; 4, front copying slide

The proposed machine had a maximum swing of 300 mm, a length between centres of 3 m and two copying slides with large tool holders, one on the front of the bed and one on the rear. Hydraulics powered the tailstock quill which had a built-in rotating centre and an axial thrust of 12,000 kgf (26, 600 lbf). The headstock was driven by a motor of 125 hp and separate feed motors were provided for each saddle, thus making more of the main motor power available for driving the work. Two other refinements were foreseen and provided for in the design: an automatic recycling

Table 9.2
Cutting data, second end of railway axle, rough turning (single-slide copying lathe)

Operation	Turned diameter (mm)	Rotational speed (rev/min)	Cutting speed (m/min)	Feed (mm/rev)	Feed (mm/min)	Radial depth of cut (mm)	Saddle travel (mm)	Cutting time (min)	Idle time (min)	Total time (min)
Cuts 1 to 5 repeated										25.8
Blend cut	162	125	64	0.3	37.5	13	94	2.4	0.6	3.1
Total for all cuts, fourth operation										28.9

device which permitted two cuts to be taken in areas where more than about 50 mm had to be removed from diameters (this would use the same template) and a special driving device for the axle. A plugboard would pro- gramme an automatic cycle.

The driving device consisted of a hydraulically operated chuck with only two jaws; the jaws, if closed under power without an axle in place, could be seen to be free to float relative to the centre carried in the head- stock spindle. The jaws were double sided so that, after gripping on the rough forged end of an axle, they could grip on the turned part for the second end by simply rotating the jaws on their locations and locking them in place, a 20 s operation. Thus, half the axle could be rough turned (the third operation) and, when the axle was lifted out of the centres, the jaws could be rotated and the axle would be immediately put back into the machine for the fourth operation because the planned cycle was always for two identical operations from the centre of the axle outwards. A full specification of the machine is given in Table 9.3; a careful examina- tion of this specification will show that the machine planned was to be approximately 26 tons instead of the normal 2 to 3 tons of a copy lathe. Whilst its maximum swing for turning purposes was 300 mm (12 in.), its bed width was 1,175 mm (46 in.), a machine which would be capable of fully utilizing its high power for long periods. The slideways were to be hardened and covered as a protection against the scale found on the out- side of the forgings and which is highly abrasive; automatic lubrication of these slideways was also provided for, in order to keep washing any scale from the surfaces, coolant would not be used as the high volume of swarf produced would frequently interrupt the flow of coolant and when it was resumed it was thought that carbide inserts in the tools would crack.

The machine tool builder provided estimated machining times with tooling lay-outs and guaranteed that 4 axles/h could be made by one machine with one operator, perhaps assisted by a labourer in loading and unloading the lathe. Ignoring loading, nine axles approximately could be machined in the time taken for one on the machine considered earlier in this chapter. There were other important factors, principally the price of the machine; however, whilst it cost four times the price of the simple machine, say, in sterling £120,000 as against £30,000, £270,000 would have to be spent on simple machines to match its output and nine operators would have to be employed instead of one. Accordingly a decision was made to go ahead with the purchase of such a machine.

The machining method for the same axle as Figure 9.2 is shown in Figure 9.4. Tools 1 and 2 are mounted on the rear copying slide and start to machine from just before the central line of the axle towards the 188 mm diameter, each of them machining half of the 162 mm diameter found on that side of the centre of the axle. Simutaneously tools 3 and 4 start to machine at one end of the axle, tool 3 making the 188 mm diameter, tool 4 roughing the 151 mm diameter; after they have traversed about 38 mm, tool 5 starts its cut to finish both the 133 and the 151 mm diameters. The

Table 9.3
Main features of automatic twin-slide copying lathe

Height of centres over the flat bed	440 mm (17½ in.)
Turning length	
maximum	3000 mm (120 in.)
minimum	1000 mm (40 in.)
Turning diameter	300 mm (12 in.)
Bed width	1175 mm (46 in.)
Spindle	
bore	100 mm (4 in.)
outside diameter in front bearing	200 mm (8 in.)
nose	15 ASA
Spindle speeds	
speeds without change of pick-off gears	8
three normal ranges	80 to 400, 50 to 250, 32 to 160 rev/min
Carriage feeds	
feeds for each carriage	11 to 58, 52 to 260, 240 to 1200 mm/min
reduction ratio on the d.c. motor	1 to 1.6
Tailstock quill	
outside diameter	200 mm (8 in.)
stroke	150 mm ($5\frac{1}{8}$ in.)
pressure on the centre	12.000 kgf (26.600 lbf)
working pressure	60 kgf/cm^2 (870 lbf/in.^{2})
Tracing slides	
cross stroke with hydraulic tracer	130 mm ($5\frac{1}{8}$ in.)
stroke with hydraulic tracer at 30°	150 mm ($5\frac{7}{8}$ in.)
thrust on the tool	3000 kgf (6.600 lbf)
working pressure	25 kgf/cm^2 (355 lbf/in.^{2})
rapid return and approach speed	10 ft/min (6.5 ft/min)
maximum tool section	60 X 40 mm
Device for automatic re-cycling	
working range	0-30 mm ($0-1\frac{3}{16}$ in.)
Hydraulic chuck	
maximum clamping diameter	300 mm (12 in.)
mimimum clamping diameter	80 mm ($3\frac{5}{32}$ in.)
working pressure	6 kgf/cm^2 (87 lbf/in.^{2})
Motors	
main motor	125 hp
d.c. motor for each carriage	3.5 hp
oil pump motor	10 hp
control motor for headstock clutches	2 hp
Net mass of the machine with two carriages (approximate)	26.000 kg (57.700 lb)

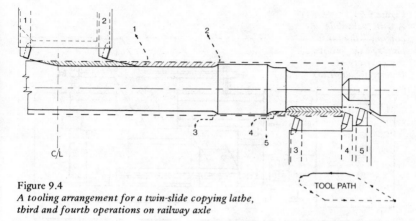

Figure 9.4
*A tooling arrangement for a twin-slide copying lathe,
third and fourth operations on railway axle*

tooling is arranged so that tools 3, 4 and 5, mounted on the front copy slide, finish cutting in approximately the same time as tools 1 and 2 on the rear slide. The cutting thrusts from the front and rear slides oppose one another and thus put little additional load on the centres supporting the axle. The tools feed into the work at 8° whilst the tools have a clearance angle of 10°.

It is interesting to note that the machine, in operation, produced a ton of swarf per hour and initially posed problems of swarf removal as its siting did not permit the use of a swarf conveyor; a large hopper was sunk into the floor and was changed periodically by crane. As this duty was performed by the labourer who assisted in loading the machine the initial calculations on machine payback were not unduly disturbed. A second machine was added later, followed by a third, the three replacing more than 25 older simpler machines. Full details of cutting data can be found in Table 9.4

CASE STUDY 2: CONSIDERATION OF TOOLING AND MACHINE FOR FINISHING

After installation of the second roughing machine, consideration was given to reducing costs incurred during the fifth operation, the finish turning of the axles. Finishing was undertaken on either single-slide copy lathes or on conventional centre lathes, the latter being employed when all the diameters were parallel such, as in the example shown earlier in this chapter.

It was decided that techniques similar to those adopted for roughing could be used for finishing, and as ceramic inserts were available for cutting tools the machine or machines purchased should be capable of making optimum use of these tools.

The final selection was a machine made by Utita in Italy which had two copying slides fitted to a near-vertical bed, hydraulic tailstock and a 40 hp motor to drive its spindle (Figure 9.5).

Figure 9.5
*A Utita twin-slide
copying lathe for
railway axle finishing
(by courtesy of
Industrial Sales Ltd)*

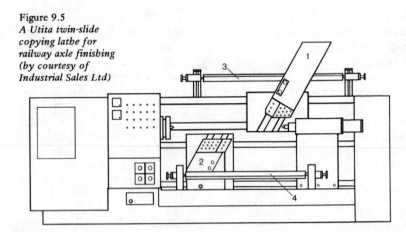

A lay-out was prepared for machining, the axle differing from the previous one as its centre section tapered from the ends towards the centre; a surface finish of 125 rms was required on all diameters but 80 rms on the radii. The lathe was able to change feed automatically whilst cutting and therefore the feed could be optimized in each of these areas. Figure 9.6

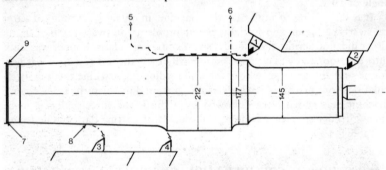

Figure 9.6
A tooling lay-put for railway axle finishing: a Utita twin-slide copying lathe

shows the tooling lay-out and the path taken by the tools; the upper tool holder carrying tools 1 and 2 starts work simultaneously with the lower tool slide whilst the chuck rotates in the opposite sense to the conventional. Tool 2 starts to cut first on the upper slide and makes the small step on the end of the axle whilst tool 1 is not in contact with the work. As soon as tool 2 commences to turn the 145 mm diameter, tool 1 also starts to cut the 212 mm diameter; as tool 1 finishes its cut, the feed is automatically reduced on the upper slide and tool 2 finishes the two radii and the 177 mm diameter on the reduced feed. The exit points for tools 1 and 2 are points 5 and 6 respectively.

The lower slide enters tools 3 and 4 at points 7 and 8 respectively and its feed is also reduced at the start of the radius at the end of the cut. It will be obvious that precise finishing is not required as the 212 mm and 145 mm diameters are to be ground afterwards but accurate tool setting is essential to provide a good blend at the point where tool 4 has started to cut and where tool 3 ends its cut, equally to leave the same grinding allowance on the 145 mm and 212 mm diameters. To facilitate this accurate tool setting tools 1 and 4 are fixed to their respective slides whilst tools 2 and 3 have adjustment slides provided with micrometer type screws. Figure 9.7 shows the lower copying slide and the tool adjustment device; the adjustable tool holder has a wide range of possible positions for tool 3 to cater for various designs of axle and the two identical knobs on the right of the slide permit two passes to be made automatically if required. The automatic feed change or end of first pass and start of second pass are controlled by trips set on drums, one for each copying slide; the drums are coupled to the two feed screws so that they make 0.8 of a complete

Figure 9.7
A view of the lower copying slide with the tool adjustment device

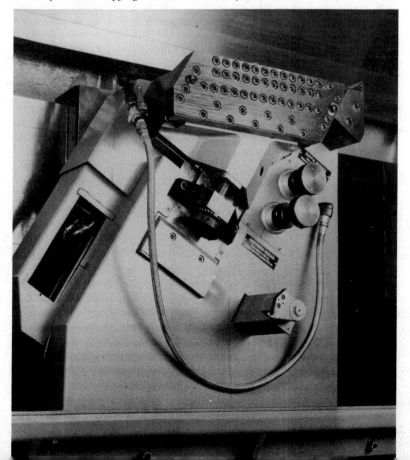

Table 9.4
Cutting data for a twin-slide copying lathe, third and fourth operations, rough turning

	Diameter (mm)	Rotational speed (rev/min)	Cutting speed (m/min)	Feed (mm/rev)	Feed (mm/min)	Number of passes	Saddle travel (mm)	Cutting time (min)	Idle time (min)	Total time (min)
Front saddle	188 (maximum)	126	74 (maximum)	0.55	70	1	355	5.00	0.40	5.40
Rear saddle	188 (maximum)	126	74 (maximum)	0.60	76	1	405	5.40	0.40	5.80

Table 9.6
Operational data, large connecting rod

Operation	Φ (mm)	Rotational speed (rev/min)	Cutting speed (m/min)	Feed (mm/rev)	Feed (mm/min)	Stroke (mm)	Cutting time (min)	Total time (min)
Unit 1, gun drill	12.2	1,800	70	0.043	77	43	0.56	
Unit 2, counterbore and chamfer	13.49	400	16.9	0.20	80	35	0.43	} 0.495
Unit 3, chamfer	13.3	400	16.9	0.20	80	5	0.065	
Unit 4, reaming	12.7	270	10	0.3	81	12	0.15	

Figure 9.8
A view showing the control drums on the right-hand end

revolution over the whole stroke of each copy slide. Figure 9.8 shows the
two drums on the extreme right of the illustration, the chuck is removed
from the machine.

The planned cutting date was as in Table 9.5.

Table 9.5
Planned cutting data

Maximum cutting speed for ceramic insert on steel of 50 to 60 kgf/mm^2	350 m/min
Machining allowance on diameters	2 to 3 mm
Feed along diameters	0.35 mm/rev
Feed around radii	0.15 mm/rev
Rapid traverse rate	7 m/min

Reference to Figure 9.6 shows that the axle is finished in two equal operations and that the largest diameter to be cut is 212 mm. The circumference of this diameter is (3.142 × 212/1000) m = 0.666 m and the required spindle speed is 350/0.66 = 525 rev/min. The nearest standard speed is 565 rev/min which is 7½% higher than the calculated ideal; as the work done by tool 1 is shorter than the work done by tool 2 and tool 2 would be cutting at lower than the optimum speed, it was felt that tool wear might be approximately equal in each case; tools 3 and 4 are also planned to run at lower than the optimum cutting speed, so tool wear, again, would not be critical.

The total cutting travel of tool 2 at the higher feed is 250 mm and the rate of tool travel is 0.35 × 565 = 198 mm/min; therefore, the time for this cut is 250/198 = 1.26 min. The total cutting travel of tool 2 at the lower feed rate is 85 mm and the rate of tool travel is 0.15 × 565 = 84.75 mm/min; the time for the cut is 85/84.75 = 1 min. The upper slide takes 2.26 min for the full pass.

The total cutting travel of the lower slide on the higher feed is 320 mm and the rate of travel is also 198 mm/min; therefore the time taken is 320 + 198 = 1.62 min. The travel of the lower slide on the reduced feed is 53 mm and the rate of travel is again 84.75 mm/min; the time taken is therefore 53 + 84.75 = 0.63 min. The lower slide takes a total of 2.25 min and is thus virtually equal to the time taken by the upper slide; rapid traverse back to the start position takes 0.15 min in each case, and the total machining time is therefore 2.41 minutes for each end of the axle.

The short machining times for rough and finish turning could easily have been badly off-set by the time taken to load and unload the axles. As a load–unload time of as little as 2 min is virtually equal to the machining time, this would have doubled the factory's finishing lathe requirements, always assuming that further machines of this type would be installed until all axles were finished by this method. A gantry loading device was installed on these machines which could handle two axles at a time, the unloading of a finished axle and insertion of the next one taking 45 s. The actual transport of the finished axle away from the machine was arranged to take place whilst the lathe was starting on the next axle. A gantry loading device for smaller parts is used; examples can be seen in Chapter 9.

CASE STUDY 3: SUPPLY OF CONNECTING RODS AND CAPS

A diesel engine manufacturer required 120 connecting rods and caps/h to supply his assembly shop. The existing plant was unsatisfactory for a number of reasons: accuracy of finished part was hard to maintain and the scrap content of the output was high; running and tooling costs were high and, as two types of connecting rod and two types of cap had to be made, the existing plant was not fully automatic; therefore labour costs were high. The problem operation was the production of bolt holes.

The accuracy problems were caused by the design of the connecting rod, demanding close-fitting bolts which acted as dowels to locate the cap with precision on the rod and very close tolerances were applied to both the hole diameters and the hole spacing. The fixtures for part location on the existing plant were made in such a way as to embody the jig bushes for the drills and reamers, and quite small variations in spacing of the bushes in the fixtures meant that many caps would not fit correctly with their rods. In later years the factory had been obliged to mark each cap and rod with a serial number and to ensure that they were each machined in the same jig–fixture. The process was slow, introduced a marking operation and did not eliminate scrap problems. Spares for engines in the field also presented a problem as persons servicing their own engines did not always realize the need to match caps and rods.

The M.S.T. company in Modena designed a special-purpose machine to make these parts which is shown in plan view in Figure 9.9.

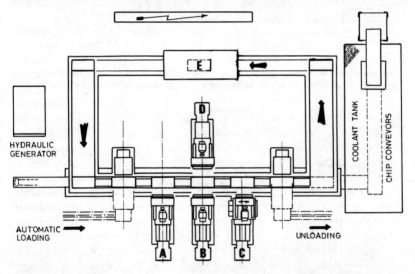

Figure 9.9
A machine for manufacturing connecting rods and caps: A, gun drilling head; B, chamfering head; C, reaming head; D, counterboring and chamfering head; E, pallet washing and drying

Four unit heads are provided to produce the bolt holes; the heads carry tools for both types of rods and caps and one set of tools is positioned below the other set on each head. The rods and caps are mounted on pallets for machining; there are two types of pallet, one for each type of part, which positions the part correctly for the appropriate spindles to do their work. The pallets are carried through the line automatically and, after the parts are unloaded, the pallets are returned to the start of the line (shown by arrows) and pass through a washing and de-swarfing machine in their passage.

Unlike the existing plant, the unit heads carry their own drill bush plates so that the same bush plates are used for every component, thus ensuring a high degree of accuracy in hole spacing. At the loading and unloading points a device is installed to clamp automatically the parts on the pallets. Each pallet carries two rods and two caps. Figure 9.10 shows an end elevation of the special-purpose machine and indicates the two working positions of the different pallets. Workpiece clamping and location positions are shown in Figure 9.11.

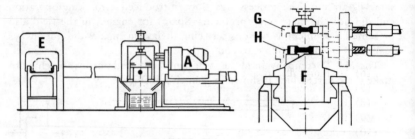

Figure 9.10
An end elevation of the machine: A, gun drilling head; E, pallet washing and drying; F, pallet; H, position of larger connecting rod; G, position of smaller connecting rod

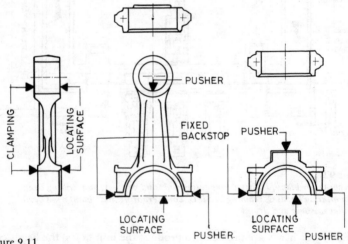

Figure 9.11
Locating points of connecting rods and caps on a pallet fixture

The sequence of operations is as follows.
(1) Load workpieces onto pallet.
(2) Hydraulically advance pallet to first unit.
(3) Gun drill holes in rod and caps (Figure 9.12).
(4) Hydraulically advance pallet to second and third units.

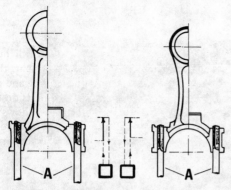

Figure 9.12
*Gun drilling holes
in rods and caps
Unit head* A

(5) With rear unit counterbore holes and chamfer and then retract the
    rear unit. Front unit, then chamfers joint face of hole (Figure
    9.13).
(6) Hydraulically advance pallet to fourth unit.
(7) Ream holes. (Action is same as first unit (Figure 9.12).)
(8) Hydraulically advance pallet to unloading station.
(9) Unload parts.
(10) Take pallets to washing station by conveyor (chain type).
(11) Wash and dry pallet.
(12) Return pallet to loading position by conveyor.

The machine had a mass of 52 tonnes complete and had a total installed
power of 200 hp. At 75% efficiency 120 rods and caps of one type are

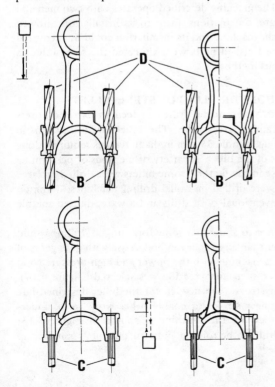

Figure 9.13
*Operations performed
by the unit heads
B, C, and D. The
operations are the
same for rods and
for caps*

Figure 9.14
A Utita deep-hole-drilling machine

machined and 110 of the larger type. The change-over from one type to the other is very short. Ancilliaries for the machine include a 2,000 litre coolant system, a hydraulic power pack and a chip conveyor.

The operational data for a large connecting rod are given in table 9.6.

Previously, nine operators were employed in performing the operations shown in Table 9.6; the results were unsatisfactory for the reasons given earlier and costs were high owing to the high labour content in the parts and the high scrap rate. The machine described operates with two men and with a negligible scrap rate. Calculations, prior to its installation, showed that the capital spent on the machine and its installation could be recovered in 3 years' running if the finished parts were allocated the same value as those made by the previous method.

CASE STUDY 4: DEEP HOLE DRILLING OF STEEL BILLETS

Nine stainless steel billets/h were to be drilled to feed an extrusion press which was producing stainless steel tube. The billets were 280 mm in diameter and 600 mm long. A hole 35 mm in diameter was required along the full length of the axis of the billet. A variety of methods of performing this operation were in use in the factory; some billets were drilled on large turret lathes and others were drilled on radial drilling machines. Common to all methods were conventional twist drills and a water-diluted soluble oil.

There was clear evidence to show that some form of gun drilling would be up to eight times faster than any other method in use with a longer tool life and lower tool costs; a machine with the power and high-pressure coolant facilities necessary for such an operation was selected (Figure 9.14). The machine has a longer stroke than needed for the billet described but was specified with the longer stroke to enable other work to be undertaken. A diagrammatic view of the machine is shown in Figure 9.15; either the billet or the drill can be rotated and a cutting oil pressure of up to 600 lbf/in.^{2} is available.

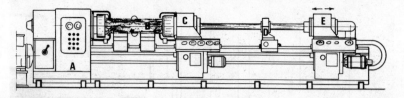

Figure 9.15
A schematic diagram of the Utita machine: A, headstock; B, workpiece; C, saddle which supports the workpiece, supports the drill, collects chips and coolant and is motorized; D, drill support; E, tailstock which feeds the drill and supplies coolant through the drill support tube

A BTA-type drill was finally selected as the best type for the work after a number of cutting tests. The cutting data is as given in Table 9.7.

Table 9.7.
Cutting data

Spindle speed	700 rev/min
Cutting speed	77 m/min
Feed/rev	0.186 mm/rev
Feed/min	130 mm/min
Machining time (approximately 600 + 130)	4.6 min

As has already been mentioned, the choice of drill type was a critical one and the choice of cutting oil was shown to be equally important as tool life and surface finish in the hole were greatly affected by the type of oil used. The tool details were as given in Table 9.8.

Table 9.8.
Tool details

Drill bit	BTA 424.6.072
Outer tube	2000 mm
Chip breaker	Code 4
Chip breaker width	1.7 mm
Chip breaker depth	0.4 mm
Radius in corner of chipbreaker	0.8 mm
Coolant pressure	20 kgf/cm^2
Coolant volume	160 litres/min
Coolant temperature (maximum)	40° C

In use this machine reduced the cost of billet drilling by 40% when a 'payback period' for the machine of 3 years was adopted. At the end of the payback period a second machine was ordered and a system of billet handling devised. Figure 9.16 shows a plan view of the two machines arranged with their headstocks adjacent and a gantry loading device to

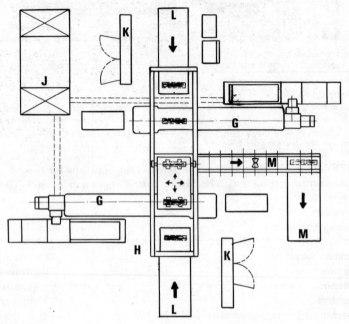

Figure 9.16
A plan view of the two machines: G, deep-bole-drilling machine; H, components; J, cooling tank; K, electrical cabinets; L, input conveyors; M, output conveyors

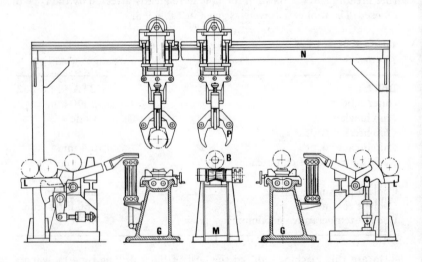

Figure 9.17
An end view of the complete installation: G, deep-bole-drilling machine; L, input conveyors, M, output conveyors; N, gantry loader; P, gripper fingers of loading arm; B, finished workpiece

take billets from input conveyors to the machines and from the machines to the single-output conveyor in the middle. A common cutting oil system was adopted for both machines which is shown in the upper right-hand corner of the diagram and one person was able to control both machines and the loading equipment.

Figure 9.17 shows an end view of the complete installation.

CASE STUDY 5: CONSIDERATION OF METHODS OF MACHINING DIESEL ENGINE

A Canadian manufacturer of large diesel engines required a replacement machine capable of finish machining 16 cylinder vee engine blocks on the external surfaces. Engine powers of up to 4000 hp are manufactured in the factory. After machining of the lower surface for a location, five other surfaces need machining; cutting must be gentle as the engine blocks are fabricated and distortion had been shown to be a problem if the surfaces were machined singly.

Plano-milling was considered but eliminated as machining all five surfaces simultaneously would have been impossible even if only on the grounds of distortion as the full width of each face along the component is machined at each pass. Planing was considered to offer the best possibility as cuts were taken along the full length (4200 mm) of the fabrication for each face and four tool boxes could be arranged to cut at the same time. Figure 9.18 shows an end view of the block with the four tools in operation

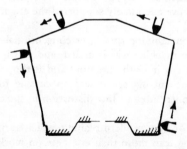

Figure 9.18
An end view of a cylinder block 4200 mm × 1270 mm × 914 mm of mass 8 tons

together, while the use of a milling head for the top surface allows the planing tools to be set at particular angles which do not have to be re-set after the milling operation.

Planing was at one time considered to be a slow operation owing to the idle time on the return stroke. Consultation with John Stirk & Sons Ltd, Halifax, produced a design of machine with tools which would cut in both forward and return directions with the tool-holder tilting at the end of each stroke.

Figure 9.19 shows the machine with the planer tool boxes set at the required angle and the milling head at the left of the cross slide. The

Figure 9.19
A planing machine with double cutting tools boxes
(by courtesy of John Stirk & Sons Ltd, Halifax)

arrangement eliminates the need for different position settings of the block and extra handling time. Alternatively, a trunnion-type fixture could have been used but would not have adequately supported the engine block except on its ends.

A 30% improvement in the machining time over conventional planing methods was achieved after the machine was installed, and this is largely due to the machine's ability to cut in both directions and at the very high speed of 107 m/min. The same cutting speed sets up cutting forces in alternate directions and thus contributes to low distortion in the cylinder block.

The machine is an example of how modern planing, which at one time was being substituted by milling, can more than compete on work of the type described, and often with a saving in tool costs when comparisons are made between the simplicity of a planer tool and a milling cutter.

10

NUMERICAL CONTROL
OF MACHINE TOOLS

Machine tools which work automatically or with little attention from an operator have been in common use since the early part of this century; until the arrival of numerical control (NC) these machines invariably took a considerable time to set for any job and in consequence were usually unsuitable for short batches. The cost of setting had to be divided amongst the parts in a batch and therefore increased the cost of making each part; on a few parts the additional cost would be substantial but on thousands of parts the reduced machining times which could be expected from automatic machining would more than off-set the very small extra cost per part of setting an automatic machine. Strenuous efforts have been made over many years to reduce setting time and thus to extend the usefulness of automatic machines to smaller and smaller batches; today, NC can sometimes be justified on batches as small as one piece done, say, four or five times per year. (An example of a computer numerical control system (CNC) is shown in Figure 10.1.)

TRIP CONTROL MOTIONS

If we first consider the conventional milling machine, a common form of automation has been to provide a single button which when pressed would start the machine spindle and would traverse the table in rapid mode up to a 'trip' which would then activate a pre-selected feed to enable cutting to commence. A second trip would stop the feed and the spindle and would return the table to the start position in rapid mode. This simple automatic cycle had (and still has) many applications but it will be obvious that it only permits one face to be machined and the cutter is not withdrawn from the work before the rapid return to start position and may thus mark the work. Figure 10.2 shows a more sophisticated form of automatic control made by Oerlikon Italiana for many years. The plate shown is mounted on the milling machine and by means of the thumbwheel switch any of the ten automatic cycles shown can be selected, thus permitting simple milling or pendulum milling cycles with two components on the table at once. The table can be lowered during the rapid return part of the cycle by operation of the lower left-hand switch on the same panel. This

Figure 10.1
*The Siemens Sinumeric 5T CNC for turning machines. The general controls can be
seen above the display area and the manual data input panel. The tape reader is inside
the hinged door at the bottom of the control*

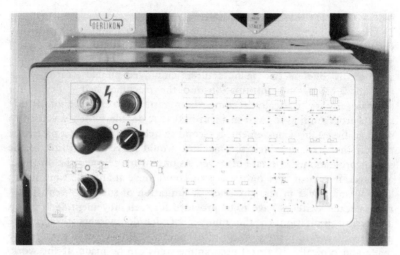

Figure 10.2
A simple automatic control giving the choice between 10 fixed cycles which are selected by the thumbwheel switch (by courtesy of Oerlikon Italiana)

latter feature was one of the first steps towards introducing a second axis of operation into milling, but it can be seen from the illustration that trips had to be set both on the table and on the vertical (knee) motion. Thus automatic machining was achieved by a combination of pre-set programmes and trips which corresponded to the dimensions of the parts being machined. This was therefore an example of 'programme control'. No cross movement of the table was provided for. (Figure 10.3.)

Figure 10.3
Trips on a milling machine table. Adjustment with precision is time consuming although 'expanding' trips are used which are quicker to set. The scale fitted to the machine table is useful as it has a vernier reader (by courtesy of Oerlikon Italiana)

DECADE SWITCH PROGRAMMING

The desire to have three-dimensional (or three-axis) machining subsequently produced a number of controls which employed plugs inserted in a matrix board, decade switches or a rotating drum to arrange the programme; additionally one or two manufacturers employed punched cards which contained programme information and the machines had a device for mechanically or optically reading these cards. An example of decade switch programming is given in Figure 10.4. A decade switch was used to select a direction of movement, a second switch used to select a particular trip on that axis to which the machine would move and a third decade switch programmed the rate of travel. As one switch selected the direction of travel, it is clear that only one axis could move at any one time and so all machining was in straight lines. A limitation of such a system is that three decade switches have to be provided for each movement and there is a physical limitation to the number of decade switches that can be provided. Therefore there is a finite number of movements that can be made and typically 30 or 60 programme steps can be made. If the work-pieces are complicated, these 30 or even 60 steps may not be sufficient; to make one cut, a rapid approach, a cutting feed and a return are needed, making three steps for a single cut. Trips have to be set as for the previous example and, as some of them have to be set with precision to ensure precision of the work, a great deal of time can be taken in setting, 60 programme steps generally needing 60 trips to be set. A further difficulty arises when a cutter becomes blunt and has to be changed; the dimensions

Figure 10.4
A decade switch type of programme control; 20 movements can be made on any axis using a feed, a creep-positioning feed or a rapid traverse. The spindle is also controlled by the switches. Trips for the cross movement can be seen in the lower left-hand corner

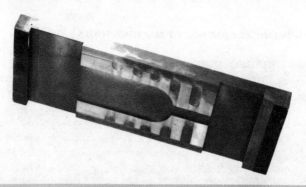

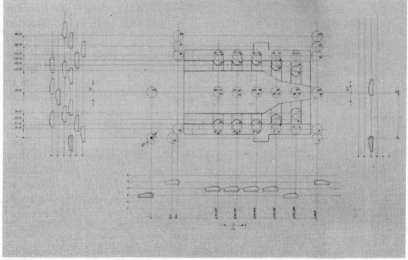

Figure 10.5
(a) *An example of a lay-out for machining an armaments part*

of the new cutter may not be precisely the same as those of the old one
and the precise trips need adjusting with consequent unproductive time
being spent.

Figure 10.5(a) shows an example of a lay-out for machining an armaments part. The actual part is shown in the upper right-hand corner of the
illustration and the trip lay-out is shown alongside the drawings of the
part. The operation is vertical milling and two parts are mounted on the
table at any time. The start position on each axis is shown by a shaded trip
and it will be noticed that some of the trips are adjustable to facilitate
setting and resetting if there is a cutter change. The second example amply
illustrates the number of trips which need to be set; a total of 24 is needed,
most of which are set with precision, to machine this heavy gun part
(Figure 10.5(b)).

Several systems have been evolved to provide programmes of nearly
infinite length and the Brown Boveri Oerlikon CPR 2 programmer is fairly
typical of them. A large belt or strip of Mylar is divided into rows in which

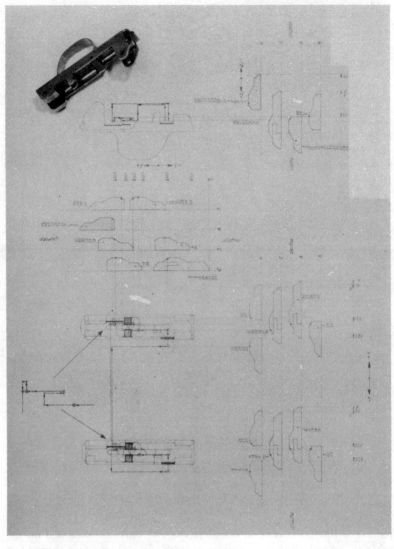

Figure 10.5
(b) Another example
illustrating the number
of trips which need to
be set

and it might be said that therein lies its greatest feature. In the course of a programme the operator can be told that he has to inspect a feature which has been machined before carrying on with the rest of the programme; this instruction, carried on the tape, will appear on the CRT display and the machine cannot continue until the operator affirms by means of the keyboard that he has carried out the instruction. The operator may ask to see the block which is being performed at a particular time or may ask to see succeeding blocks of information; he may ask for a display of the distance yet to be travelled on an axis or to see the values of stored tool error corrections.

Mention has already been made of the facility to display parts of the programme and to edit them and there are other messages which appear on the CRT display automatically; these include warning messages such as 'illegal tape format' where, say, a block has been programmed incorrectly or fault messages to tell the operator of a fault in the machine or system. Comprehensive diagnostic facilities are frequently available with a CNC system, including a display of logic ladder diagrams which enables faulty machine parts such as micro-switches or pressure valves to be found rapidly (Figure 10.19).

MILLING OF IMPELLER BY COMPUTER NUMERICAL CONTROL

As has already been stated, a deeper study of any one CNC system is beyond the scope of this chapter, as is also detailed programming information, but the following example will give, in some detail, an impression

Figure 10.19
A display of part of the interface ladder diagram for a machine tool on the Plessey 7360 CNC. As each element in the ladder diagram is activated, it is illuminated to double brilliance on the screen, thus confirming that each function occurs

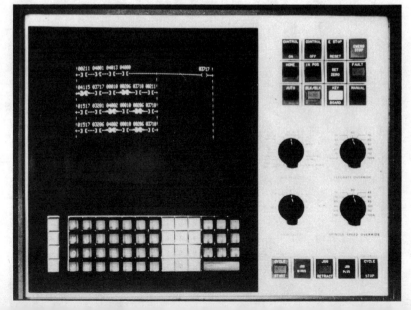

Figure 10.20
A turned blank and the finished dimpeller machined by a CNC milling machine (by courtesy of Oerlikon Italiana)

Figure 10.21
The indexing fixture for the impeller mounted on the machine table. A swarf conveyor is mounted below the fixture to clear the large amounts of swarf made during milling and to ensure that machining does not have to stop because of an accumulation

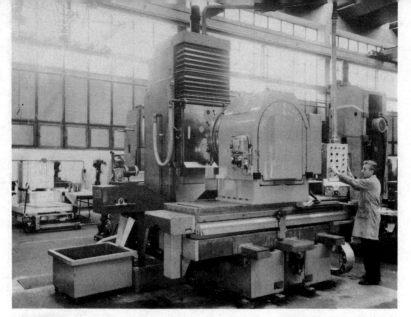

Figure 10.22
The Oerlikon FB4 CNC vertical–horizontal milling machine used to machine the impeller

of the capabilities of CNC milling.

The workpiece is a high-speed stainless steel impeller which must be of a high degree of precision. A large powerful milling machine was selected for the work and was equipped with an indexing fixture to reduce programming complexity. Figure 10.20 shows the turned blank prior to milling and the part after all milling is completed. The indexing fixture seen in Figure 10.21 permits one slot to be programmed, milled and repeated seventeen times after indexing to finish the part. The complete machine is shown in Figure 10.22 and its control system also controls the indexing of the fixture. Five cuts are taken to produce each slot, using different cutters for each slot; the cutter paths are shown in Figure 10.23.

The first operation uses a 'ripper' type cutter 63 mm in diameter which is operated at 355 rev/min, giving a cutting speed of 70 m/min. For the cut S_1 a slower feed has to be used than for S_2, as on the first cut the tool is entering solid metal. The subsequent operations and their machining times are also shown. In this particular case cutter changes are made manually with the aid of a power drawbar and the cutters used are shown in Figure 10.24. The first operation is illustrated in Figure 10.25; the multi-point clamping of the part is clearly shown. (An example of the copy milling of another impeller is given in Chapter 4.)

As the programmes for NC or CNC systems are written with nominal tool dimensions used at the planning stage, tool corrections in the form of dimensions corresponding to the difference between the theoretical position of each tool and the actual position are applied when the programme is used. For milling and drilling–boring machines these compensations are made for variations in both diameter and length of the tool, and

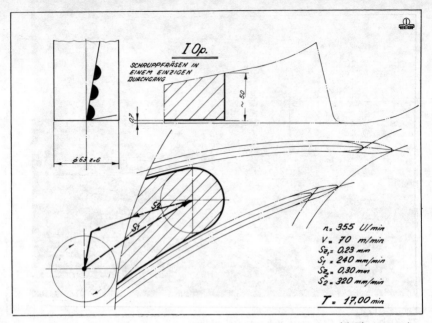

Figure 10.23   *(a) First operation*

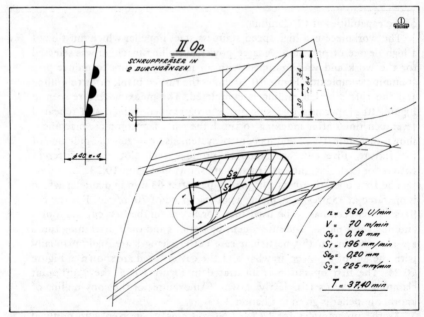

(b) Second operation

for turning machines the compensations are applied for mispositioning of the tools in the tool holders in two planes corresponding to the two axes of the machine.

Some early NC systems had a very restricted capacity for tool correc-

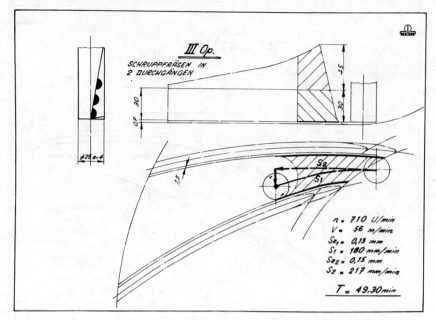

III Op.

SCHRUPPFRÄSEN IN
2 DURCHGÄNGEN

$n = 710$ U/min
$V = 56$ m/min
$S_{z_1} = 0,13$ mm
$S_1 = 180$ mm/min
$S_{z_2} = 0,15$ mm
$S_2 = 217$ mm/min

$T = 49,30$ min

(c) *Third operation*

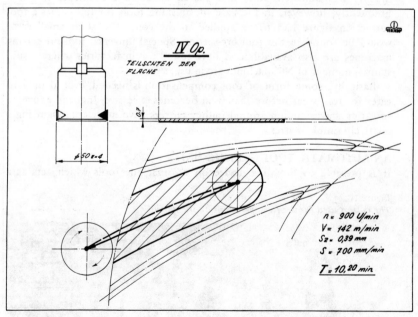

IV Op.

TEILSCHTEN DER
FLÄCHE

$n = 900$ U/min
$V = 142$ m/min
$S_z = 0,39$ mm
$S = 700$ mm/min

$T = 10,20$ min

(d) *Fourth operation*

tions; within the system mispositioning of up to ±0.025 in. on each axis could be catered for and tool pre-setting facilities were usually provided by the user of the machine. Reasonably simple devices to hold a tool-holder in the same manner as it was held in the machine were equipped with dial

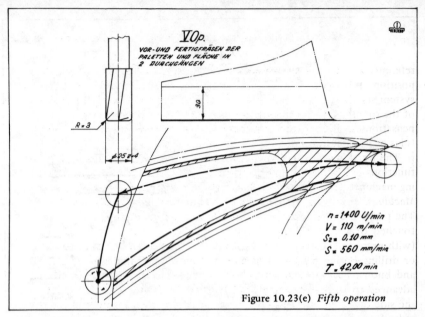

VOp.
VOR- UND FERTIGFRÄSEN DER
PALETTEN UND FLÄCHE IN
2 DURCHGÄNGEN

30

R = 3

ø 25 z 4

$n = 1400 \ U/min$
$V = 110 \ m/min$
$S_z = 0,10 \ mm$
$S = 560 \ mm/min$
$T = 42,00 \ min$

Figure 10.23(e) *Fifth operation*

indicators or some other means of measuring the tool point position and
thus it was adjusted until it appeared to be in the correct position.
Frequently, however, in the case of finishing tools for fine limits a tool
correction figure had to be applied to the control as some small error
would be found in the tool pre-setting. Special universal tool pre-setting
machines are also available and have found use in factories where a substantial number of NC machines are in use.

Basically, some form of tool compensation is needed, even if only to
cater for tool wear before that wear becomes excessive. Indexable turning
tool tips and precision ground milling cutters have made tool changing a
relatively simple matter.

AN AUTOMATIC TOOL SETTING UNIT

It is possible to mount a device on NC machine tools which acts as a

Figure 10.24
The cutters used for the five operations on the impeller

reference datum and against which all the tools are compared for their positions relative to the theoretical position. When coupled to a CNC system such a unit can, with the appropriate software, check the position of the tool in two axes and can automatically insert the required tool compensation dimensions, the tools only having been pre-set to ruler accuracy (plus or minus a few millimetres).

The Renishaw TP1/T Touch Trigger Probe is shown in Figure 10.26 and functions in a similar manner to those used on three-dimensional measuring machines. (For further details, see H.C. Town and H. Moore, *Inspection Machines, Measuring Systems and Instruments*, Batsford, London, 1978.) The probe is spring loaded to return always to its rest position and can be touched from any direction including axially to send an electrical signal (with extreme accuracy of repeatability) to the CNC system. For a milling or drilling–boring machine it is mounted on the machine table in a fixed and known position and a new tool, mounted in the machine spindle, is advanced towards it; the flute of the cutter are made to touch the flat on the side of the probe. This operation is performed on a slow feed and the pulse from the touch trigger probe is used to stop the feed and to instruct the CNC to measure the position of the table at the moment that the signal was sent. If a milling cutter had been reground several times and its diameter was 0.010 in. smaller than nominal, the machine would have

Figure 10.25
The part machined at the first operation on the impeller

traversed 0.005 in. past the point at which the tool should have touched the probe. This dimension of 0.005 in. is recorded by the CNC unit and is used as a tool error compensation for that particular tool.

Similarly the end of the tool can then be touched on the top of the probe and any error in the length of the tool from the spindle nose will also be recorded and used as a compensation. Thus tool error compensation can be performed automatically and tool pre-setting with accuracy an unnecessary chore.

FINISHED-PART INSPECTION ON THE MACHINE

A similar probe can be fitted iwth a taper shank and can be kept in the tool magazine of either a turning machine or a machining centre equipped with an automatic tool changer. When a workpiece is completed, the last cutting tool can be returned to the magazine and the measuring probe can

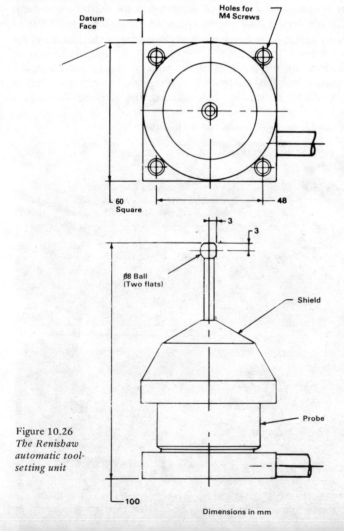

Figure 10.26
The Renishaw automatic tool-setting unit

Dimensions in mm

be inserted in the machine spindle. Electrical coupling to the CNC is achieved inductively through a coil mounted in the unit and a coil mounted near or on the spindle nose so that signal wires do not need to be connected. A programme for the inspection cycle is included at the end of the machining programme and the probe is taken round the finished part by the machine, touching the various machined surfaces and comparing their position with the theoretical positions; tolerance information can also be included. Hence the machine which makes the part can also check its work. Figure 10.27 shows such a probe mounted in the spindle of a machining centre.

The 1980s are expected to bring an increase in the use of direct numerical control (DNC) where a large number of machine tools are controlled by one central computer which plans and controls the shop loading and routing of parts and which also arranges inspection of the various products of the factory. Because all programme information is stored centrally with merely a terminal on each controlled machine, this should provide a high degree of flexibility with optimum use of the factory facilities.

Figure 10.27
The Renishaw touch trigger finger mounted in the spindle of a CNC machining centre prior to measuring the dimensions of a completed workpiece

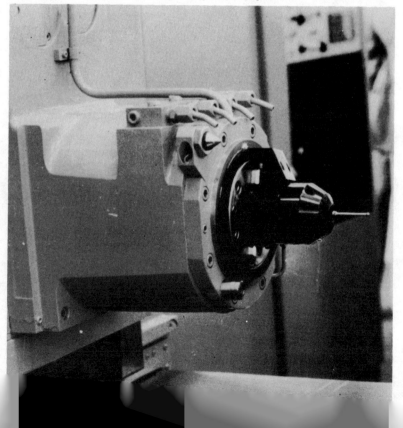

MACHINE TOOL LABORATORY WORK

The examples given in this chapter include machining operations and the calculations required to produce the more intricate profiles in the specialized field of engineering developments. In many instances they provide a means of extending the normal use of machine tools found in educational establishments into the realms of laboratory, experimental and tool room work for the more advanced students.

PROBLEMS OF UNIVERSAL MILLING

EXAMPLE 1: DIFFERENTIAL INDEXING

Figure 11.1 shows the mechanism of a dividing head arranged for differential indexing. Fifteen change gears are supplied, with 24, 28, 32, 40, 44, 48, 56, 64, 68, 72, 80, 84, 86, 96 and 100 teeth.

With a selection of these gears connected from the shaft actuating the index plate to the main spindle, indexing can be carried out for numbers of divisions not available by simple indexing. In the latter case, the index

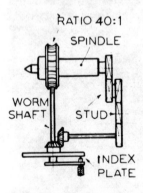

Figure 11.1
The gear arrangement for differential indexing

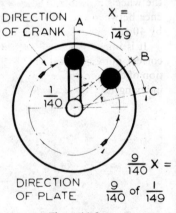

Figure 11.2
The index plate and crank movement

plate is locked against movement; however, for differential indexing it must be free to revolve for, when the crank is turned, the spindle revolves through the worm and wheel, and the index plate is moved by gearing in either the same or the opposite direction to that of the crank handle.

Assume that 149 divisions are required; arrange the indexing for some simple number, say, 140, and then provide it with some additional movement by the gear train which will automatically subtract the difference between 1/140 and 1/149 from the movement of the crank, resulting in an effective movement of 1/149.

Figure 11.2 shows the movement of the crank and plate, A and C representing the angular movement of the crank and the hole in the plate at the beginning and during the movement the hole approaches the crank. When the movement has terminated, it has consisted of two parts, AB owing to the crank and BC owing to the plate.. The difference between 1/140 and 1/149 is 149 − 140/140 × 149 = 9/140 of 1/149. Therefore the ratio between the movements of crank and plate is 9/140 and BC is thus 9/140 of AB.

To show the algebraic relation, let x be the movement by crank AB. Then (9/140) x is the movement made by the index plate in the same time, and this is equal to BC. The two movements combined must equal 1/140:

$$x + (9/140)\,x = 1/140$$
$$140x + 9x = 1$$
$$149x = 1$$
$$x = 1/149.$$

The ratio of 9/140 for the gear train includes the constant ratio of 1 to 40 for the worm and wheel. Dividing 9/140 by 1/40 gives the ratio for the change wheels only; thus 9/140 × 40/1 = 360/140 = 18/7 or 72/28 from the wheels available. The rule for the change wheel ratio is that the difference between the number required and the number indexed for multipled by 40 and divided by the number indexed for.

It is important when mounting the change wheels to take care that the correct number of intermediates are used; otherwise the direction of rotation of the index plate may not be correct. For example, to cut 127 spaces, indexing for 120 gives a difference of 7, but 113 spaces can be cut just as

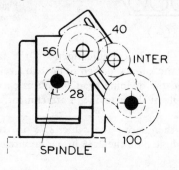

Figure 11.3
*The end view of
a dividing head*

easily by a change in the number of intermediates used. From the formula given, to cut 201 divisions would require a ratio of (201 − 200) × 40/200 = 4/20 = 1/2 × 4/10 or as gears 28/56 × 40/100, these being mounted as in Figure 11.3 using one intermediate.

EXAMPLE 2: PROBLEM IN INDEXING

The operation is that of boring 19 holes in the cast iron jig plate Figure 11.4(a), the holes being spaced as shown. Two set-ups are feasible: (1) with the dividing head set (Figure 11.4(b)) paralleled to the machine spindle; (2) with the dividing head spindle vertical and the tool held in the vertical milling attachment. In case (1) the feed to the table is in a transverse

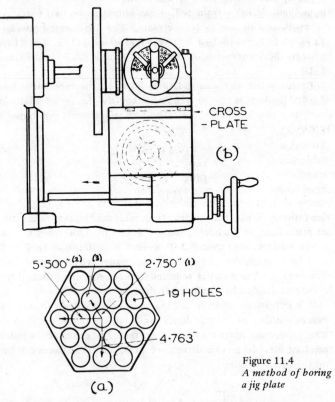

Figure 11.4
*A method of boring
a jig plate*

direction, and in case (2) by vertical adjustment of the table. The first operation is to rough drill the holes and then to bore to size. Concentricity between centre hole 1 and the dividing head spindle is checked by a dial indicator held on the spindle of the machine or attachment.

Procedure

Bore centre hole 1 and then move the table 2.75 in. to the right to position the plate to bore hole 2 on the 2.75 in. radius circle. Position hole

3 on the same circle by indexing 40/6 = 6 turns and 16 spaces on the 24-hole circle of the index plate. When the six holes on this circle have been machined, the machine table is moved to the right 4.763 − 2.750 = 2.013 in. This locates the plate for machining the six holes on the 4.763 in. radius circle.

The six holes in this circle are spaced equally, but are displaced 30° with respect to the holes in the previous circle. Hence, to position hole 8 (the first hole on the 4.763 radius circle) in line with the tool, it is now necessary to index the plate 30° or the equivalent amount corresponding to 12 divisions, ie 40/12 = 3 turns and 8 additional spaces on the 24-hole circle. Hole 9 and each of the remaining holes are indexed by turning the crank 6 turns and 16 additional holes on the 24-hole circle, the same indexing as used on the 2.75 radius circle.

The same procedure is used when setting the place for machining holes 14 to 19 inclusive on the 5.5 in. radius.

Note that the indexing for hole 8, displaced at 30° from the previous hole, could have been obtained from the following: the angular indexing is 360/40 = 9° or 1 degree = 1/9 turn of indexing crank; therefore indexing for 30° gives 30/9 or $3\frac{1}{3}$, $3\frac{8}{24}$ as given when using a 24-hole circle.

EXAMPLE 3: HELICAL MILLING

To cut a helix on a milling machine, the dividing head is geared to the table screw so that, as the table is traversed, a rotary motion is given to the spindle of the dividing head. If equal wheels are mounted on the dividing head and the table screw a pitch of $\frac{1}{5}$ in. pitch, because 40 turns of the worm shaft of the head are required for one revolution of the spindle, the lead obtained is 40/5 = 8 in. (Note that this is 10 in. where the screw is ¼ in pitch.)

To find the gear train for any helix, the formula is

$$\frac{\text{lead of helix to be cut}}{\text{lead of machine 8 in.}} = \frac{\text{product of driven gears}}{\text{product of driving gears}}$$

Procedure

The procedure to cut a right-hand spiral gear with 24 teeth, a diametral pitch of 10 and an angle of 28.5° is as follows. The lead of the spiral, based on the pitch line diameter, can be found from

tangent of angle = circumference of work ÷
lead of spiral 0.543 = 3.14 × 2.2/lead = 12.722 in. lead

For a right-hand spiral, the right-hand end of the table is pushed towards the column of the machine, as shown in Figure 11.5, but this should not be done until the cutter has been set over the centre of the blank.

The gear train to give a lead of 12.722, from the formula given, is thus 12.722/8 or, from the wheels supplied, 32 × 48/ 40 × 24 or a similar ratio. These are mounted as shown in Figure 11.6.

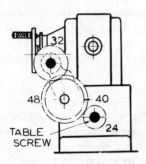

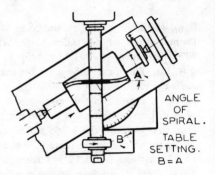

Figure 11.6
A set-up for milling spiral gear

Figure 11.5
*The mounting of gears for cutting
teeth in a blank*

Spur gear cutters for milling helical gears are selected for the hypothetical number of teeth in the section at right angles to the helix rather than the actual number of teeth to be cut, and the formula used is

$$\frac{\text{number of teeth in wheel}}{\cos^3 C}$$

where C is the helix angle in degrees.

Having got the machine set up, the depth of tooth to be cut is obtained from 2.157/ diametral pitch = 0.216 in., while the index sectors for dividing for the 24 teeth require to be set to give 40/24 or $1^{16}/_{24}$ turns for each tooth.

After taking the first cut to about 0.180 in deep, return the cutter clear of the blank, and index for the next tooth, so continuing around the blank. Then sink the cutter to full depth, and repeat the operation. It is essential when spiral milling that a blank is tight on a mandrel, for the twisting action of the cut tends to slacken the work. Also, cutting should always be done towards the dividing head and not towards the tailstock.

EXAMPLE 4: FLUTING AN END MILL

For cutting the flutes of helical teeth end mills or twist drills and reamers, if special-shaped cutters are available the normal set-up is by mounting the blank between the dividing head and tailstock centres. If, however, special cutters are not available, a vertical spindle attachment can often be used with a standard cutter, as shown in Figure 11.7. Other helical work carried out in this manner includes the milling of helical splines or conical spirals on tapered milling cutters.

The example given is a six-fluted end mill, 1 in. in outside diameter, right-hand cutting with a 12° left-hand spiral and a flute angle of 95° with a corner radius of 0.031 in. The blank is mounted as shown and a 0.525 in. end mill with a 0.031 in. radius on the end teeth is held in the spindle of the vertical head.

The lead of spiral is tan (12°) = 3.14 × 1/ lead = 0.212 = 3.14 /lead or

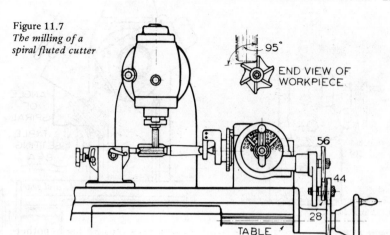

Figure 11.7
*The milling of a
spiral fluted cutter*

95°

END VIEW OF
WORKPIECE

56

44

28

TABLE
SCREW   48

the lead = 14.81 in., giving a required gear ratio of 14.81/8 (lead of
machine) = 56 × 44 / 28 × 48 or a similar ratio mounted to connect the
table screw and dividing head as shown.

As there are six flutes, the sectors on the dividing head index plate will
be set to divide for each flute in turn, ie 40/6 = $6^2/_3$ turns, or 6 complete
turns of the crank + 24 holes on the 36-hole circle. (Note that the 24 holes
are in addition to the hole used for the index pin.)

The cutter is off-set from the blank centre by $^5/_{16}$ in. as shown in the
end view of the work and, as left-hand flutes are required, the left-hand
end of the table must be pushed away when the operator is facing the
front of the machine and then must clamp at the 12° setting.

Procedure

As the cutter has a flat end, two cuts are required to produce the 95° angle,
the first one giving 90° only when the cutter is set to depth and the first
traverse made. The workpiece is then indexed for the next flute, and the
operation is continued for all six flutes.

To remove the last 5° of metal to obtain the shape required, the blank
is indexed by the crank handle 360/40 = 9 or as 1° = 1/9 turn; then 5/9
turn are required to bring the blank into position for the final cut; as the
sectors are already set in the 36-hole circle, it is a simple matter to index
20 holes in this circle to locate the cutter to give the required angle of 95°.
Thus, by machining and indexing as before for each of the flutes, the work
is completed.

Assuming that the blank is of high-speed steel, annealed to 200 Brinell,
a cutting speed of 150 ft/min is suitable, with a feed per tooth of 0.002 in.
on the cutter in the vertical head. H.S.S. cutter.

EXAMPLE 5: GRADUATING (1)

The milling machine equipped with an indexing head can be used for
graduating flat scales, verniers and other parts requiring odd fractional

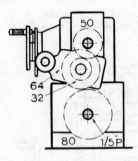

Figure 11.8
A gear arrangement for graduating

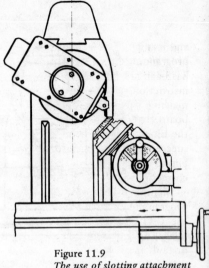

Figure 11.9
*The use of slotting attachment
for graduating*

divisions or graduations. The spindle of the dividing head must be geared to the table feed screw, as in Figure 11.8, so that a longitudinal movement of the table is secured by turning the index crank. The driving gear is mounted on an arbor in the bore of the spindle, in a similar manner to when it is used for differential indexing.

By varying the indexing movement, graduations can be spaced with considerable accuracy. The graduation lines can be cut by a single-point tool held either in a fly cutter arbor mounted in the main spindle of the machine or between collars on the milling cutter arbor. The lines are produced by feeding the table transversely by hand, and the length of lines representing various divisions and subdivisions can be varied by noting the graduations on the cross feed screw.

If the gearing between the dividing head spindle and the table screw is equal, then 1 turn of the index crank will give a longitudinal table traverse movement of $1/40 \times 1/5$ (table screw) = 0.005 or 1/200 in. It is thus easy by utilizing the index plates to graduate any numbers divisible by five but, assuming that graduating lines 0.03125 or 1/32 in apart are required to graduate a scale, then the change wheels can be employed to vary the table traverse per one revolution of the index crank.

Thus for the dividing head spindle and table screw to rotate at the same speed the gear ratio should be 5 to 4, the pitch of the table screw being 1/5 in., and must be speeded up. The ratio $1/40 \times 5/4 = 5/160 = 0.03125$ in., suitable change wheels being $50/32 \times 64/80$ as shown in the diagram.

Thus the number of turns of the index crank for moving the table 1/32 in = 0.03125/0.00625 = 5 turns, the figure of 0.00625 being obtained from the total ratio of $1/40 \times 5/4 \times 1/5 = 0.00625$.

If the divisions on a vernier reading to thousandths of an inch were to be 0.024 in. apart, the indexing movement would equal 0.024/0.00625 = 3.84 turns. This fractional movement of 0.84 turns can be obtained within and it might be said that therein lies its greatest feature. In the course of a programme the operator can be told that he has to inspect a feature which has been machined before carrying on with the rest of the programme; this instruction, carried on the tape, will appear on the CRT display and the machine cannot continue until the operator affirms by means of the keyboard that he has carried out the instruction. The operator may ask to see the block which is being performed at a particular time or may ask to see succeeding blocks of information; he may ask for a display of the distance yet to be travelled on an axis or to see the values of stored tool error corrections.

Mention has already been made of the facility to display parts of the programme and to edit them and there are other messages which appear on the CRT display automatically; these include warning messages such as 'illegal tape format' where, say, a block has been programmed incorrectly or fault messages to tell the operator of a fault in the machine or system. Comprehensive diagnostic facilities are frequently available with a CNC system, including a display of logic ladder diagrams which enables faulty machine parts such as micro-switches or pressure valves to be found rapidly (Figure 10.19).

MILLING OF IMPELLER BY COMPUTER NUMERICAL CONTROL

As has already been stated, a deeper study of any one CNC system is beyond the scope of this chapter, as is also detailed programming information, but the following example will give, in some detail, an impression

Figure 10.19
A display of part of the interface ladder diagram for a machine tool on the Plessey 7360 CNC. As each element in the ladder diagram is activated, it is illuminated to double brilliance on the screen, thus confirming that each function occurs

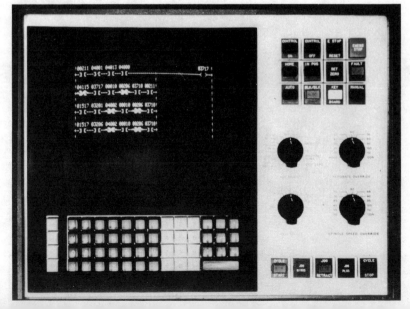

Figure 10.20
A turned blank and the finished dimpeller machined by a CNC milling machine (by courtesy of Oerlikon Italiana)

Figure 10.21
The indexing fixture for the impeller mounted on the machine table. A swarf conveyor is mounted below the fixture to clear the large amounts of swarf made during milling and to ensure that machining does not have to stop because of an accumulation

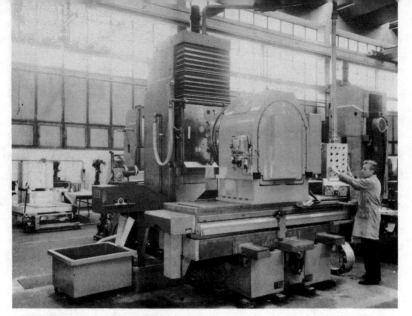

Figure 10.22
The Oerlikon FB4 CNC vertical–horizontal milling machine used to machine the impeller

of the capabilities of CNC milling.

The workpiece is a high-speed stainless steel impeller which must be of a high degree of precision. A large powerful milling machine was selected for the work and was equipped with an indexing fixture to reduce programming complexity. Figure 10.20 shows the turned blank prior to milling and the part after all milling is completed. The indexing fixture seen in Figure 10.21 permits one slot to be programmed, milled and repeated seventeen times after indexing to finish the part. The complete machine is shown in Figure 10.22 and its control system also controls the indexing of the fixture. Five cuts are taken to produce each slot, using different cutters for each slot; the cutter paths are shown in Figure 10.23.

The first operation uses a 'ripper' type cutter 63 mm in diameter which is operated at 355 rev/min, giving a cutting speed of 70 m/min. For the cut S_1 a slower feed has to be used than for S_2, as on the first cut the tool is entering solid metal. The subsequent operations and their machining times are also shown. In this particular case cutter changes are made manually with the aid of a power drawbar and the cutters used are shown in Figure 10.24. The first operation is illustrated in Figure 10.25; the multi-point clamping of the part is clearly shown. (An example of the copy milling of another impeller is given in Chapter 4.)

As the programmes for NC or CNC systems are written with nominal tool dimensions used at the planning stage, tool corrections in the form of dimensions corresponding to the difference between the theoretical position of each tool and the actual position are applied when the programme is used. For milling and drilling–boring machines these compensations are made for variations in both diameter and length of the tool, and

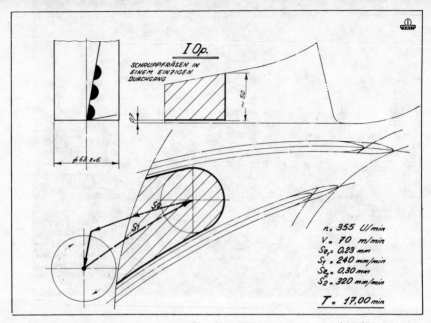

Figure 10.23

(a) *First operation*

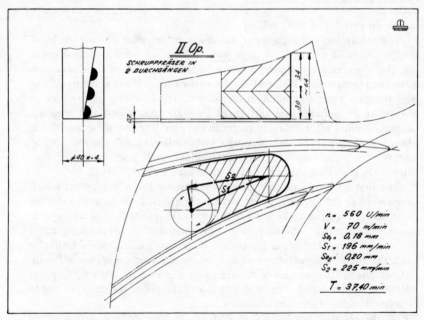

(b) *Second operation*

for turning machines the compensations are applied for mispositioning of the tools in the tool holders in two planes corresponding to the two axes of the machine.

Some early NC systems had a very restricted capacity for tool correc-

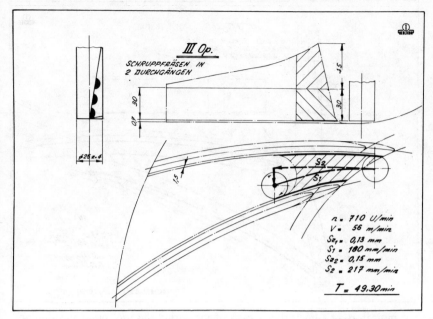

(c) *Third operation*

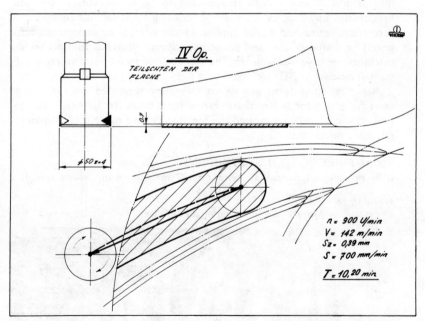

(d) *Fourth operation*

tions; within the system mispositioning of up to ±0.025 in. on each axis
could be catered for and tool pre-setting facilities were usually provided by
the user of the machine. Reasonably simple devices to hold a tool-holder
in the same manner as it was held in the machine were equipped with dial

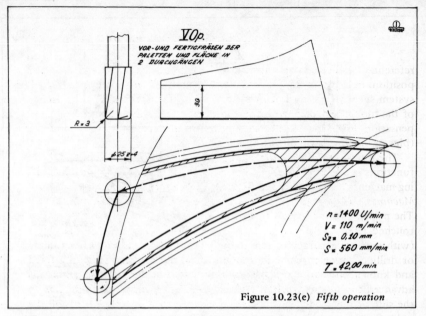

VOp.

VOR- UND FERTIGFRÄSEN DER
PALETTEN UND FLÄCHE IN
2 DURCHGÄNGEN

30

R = 3

6,25 z = 4

$n = 1400 \text{ U/min}$
$V = 110 \text{ m/min}$
$S_z = 0,10 \text{ mm}$
$S = 560 \text{ mm/min}$
$T = 42,00 \text{ min}$

Figure 10.23(e) *Fifth operation*

indicators or some other means of measuring the tool point position and thus it was adjusted until it appeared to be in the correct position. Frequently, however, in the case of finishing tools for fine limits a tool correction figure had to be applied to the control as some small error would be found in the tool pre-setting. Special universal tool pre-setting machines are also available and have found use in factories where a substantial number of NC machines are in use.

Basically, some form of tool compensation is needed, even if only to cater for tool wear before that wear becomes excessive. Indexable turning tool tips and precision ground milling cutters have made tool changing a relatively simple matter.

AN AUTOMATIC TOOL SETTING UNIT
It is possible to mount a device on NC machine tools which acts as a

Figure 10.24
The cutters used for the five operations on the impeller

reference datum and against which all the tools are compared for their positions relative to the theoretical position. When coupled to a CNC system such a unit can, with the appropriate software, check the position of the tool in two axes and can automatically insert the required tool compensation dimensions, the tools only having been pre-set to ruler accuracy (plus or minus a few millimetres).

The Renishaw TP1/T Touch Trigger Probe is shown in Figure 10.26 and functions in a similar manner to those used on three-dimensional measuring machines. (For further details, see H.C. Town and H. Moore, *Inspection Machines, Measuring Systems and Instruments*, Batsford, London, 1978.) The probe is spring loaded to return always to its rest position and can be touched from any direction including axially to send an electrical signal (with extreme accuracy of repeatability) to the CNC system. For a milling or drilling–boring machine it is mounted on the machine table in a fixed and known position and a new tool, mounted in the machine spindle, is advanced towards it; the flute of the cutter are made to touch the flat on the side of the probe. This operation is performed on a slow feed and the pulse from the touch trigger probe is used to stop the feed and to instruct the CNC to measure the position of the table at the moment that the signal was sent. If a milling cutter had been reground several times and its diameter was 0.010 in. smaller than nominal, the machine would have

Figure 10.25
The part machined at the first operation on the impeller

traversed 0.005 in. past the point at which the tool should have touched the probe. This dimension of 0.005 in. is recorded by the CNC unit and is used as a tool error compensation for that particular tool.

Similarly the end of the tool can then be touched on the top of the probe and any error in the length of the tool from the spindle nose will also be recorded and used as a compensation. Thus tool error compensation can be performed automatically and tool pre-setting with accuracy an unnecessary chore.

FINISHED-PART INSPECTION ON THE MACHINE

A similar probe can be fitted iwth a taper shank and can be kept in the tool magazine of either a turning machine or a machining centre equipped with an automatic tool changer. When a workpiece is completed, the last cutting tool can be returned to the magazine and the measuring probe can

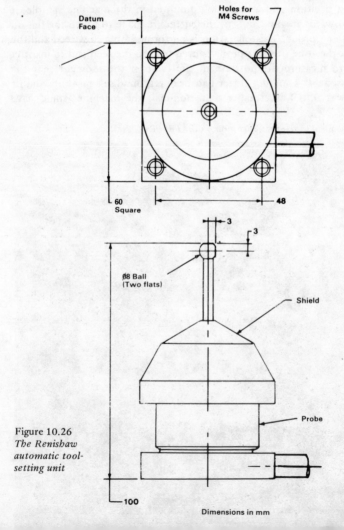

Figure 10.26
The Renishaw automatic tool-setting unit

Dimensions in mm

be inserted in the machine spindle. Electrical coupling to the CNC is achieved inductively through a coil mounted in the unit and a coil mounted near or on the spindle nose so that signal wires do not need to be connected. A programme for the inspection cycle is included at the end of the machining programme and the probe is taken round the finished part by the machine, touching the various machined surfaces and comparing their position with the theoretical positions; tolerance information can also be included. Hence the machine which makes the part can also check its work. Figure 10.27 shows such a probe mounted in the spindle of a machining centre.

The 1980s are expected to bring an increase in the use of direct numerical control (DNC) where a large number of machine tools are controlled by one central computer which plans and controls the shop loading and routing of parts and which also arranges inspection of the various products of the factory. Because all programme information is stored centrally with merely a terminal on each controlled machine, this should provide a high degree of flexibility with optimum use of the factory facilities.

Figure 10.27
The Renishaw touch trigger finger mounted in the spindle of a CNC machining centre prior to measuring the dimensions of a completed workpiece

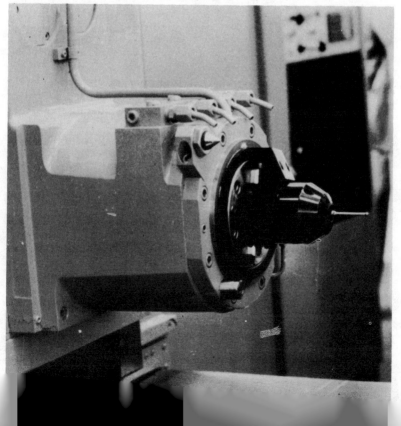

11

MACHINE TOOL
LABORATORY WORK

The examples given in this chapter include machining operations and the
calculations required to produce the more intricate profiles in the
specialized field of engineering developments. In many instances they provide a means of extending the normal use of machine tools found in educational establishments into the realms of laboratory, experimental and tool
room work for the more advanced students.

PROBLEMS OF UNIVERSAL MILLING

EXAMPLE 1: DIFFERENTIAL INDEXING
Figure 11.1 shows the mechanism of a dividing head arranged for differential indexing. Fifteen change gears are supplied, with 24, 28, 32, 40, 44,
48, 56, 64, 68, 72, 80, 84, 86, 96 and 100 teeth.

With a selection of these gears connected from the shaft actuating the
index plate to the main spindle, indexing can be carried out for numbers
of divisions not available by simple indexing. In the latter case, the index

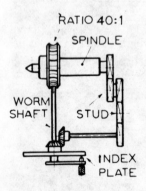

Figure 11.1
*The gear arrangement for
differential indexing*

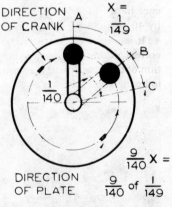

Figure 11.2
*The index plate and crank
movement*

plate is locked against movement; however, for differential indexing it must be free to revolve for, when the crank is turned, the spindle revolves through the worm and wheel, and the index plate is moved by gearing in either the same or the opposite direction to that of the crank handle.

Assume that 149 divisions are required; arrange the indexing for some simple number, say, 140, and then provide it with some additional movement by the gear train which will automatically subtract the difference between 1/140 and 1/149 from the movement of the crank, resulting in an effective movement of 1/149.

Figure 11.2 shows the movement of the crank and plate, A and C representing the angular movement of the crank and the hole in the plate at the beginning and during the movement the hole approaches the crank. When the movement has terminated, it has consisted of two parts, AB owing to the crank and BC owing to the plate.. The difference between 1/140 and 1/149 is 149 − 140/140 × 149 = 9/140 of 1/149. Therefore the ratio between the movements of crank and plate is 9/140 and BC is thus 9/140 of AB.

To show the algebraic relation, let x be the movement by crank AB. Then $(9/140)\, x$ is the movement made by the index plate in the same time, and this is equal to BC. The two movements combined must equal 1/140:

$$x + (9/140)\, x = 1/140$$
$$140x + 9x = 1$$
$$149x = 1$$
$$x = 1/149.$$

The ratio of 9/140 for the gear train includes the constant ratio of 1 to 40 for the worm and wheel. Dividing 9/140 by 1/40 gives the ratio for the change wheels only; thus 9/140 × 40/1 = 360/140 = 18/7 or 72/28 from the wheels available. The rule for the change wheel ratio is that the difference between the number required and the number indexed for multipled by 40 and divided by the number indexed for.

It is important when mounting the change wheels to take care that the correct number of intermediates are used; otherwise the direction of rotation of the index plate may not be correct. For example, to cut 127 spaces, indexing for 120 gives a difference of 7, but 113 spaces can be cut just as

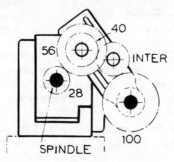

Figure 11.3
The end view of a dividing head

easily by a change in the number of intermediates used. From the formula given, to cut 201 divisions would require a ratio of $(201 - 200) \times 40/200$ = $4/20 = 1/2 \times 4/10$ or as gears $28/56 \times 40/100$, these being mounted as in Figure 11.3 using one intermediate.

EXAMPLE 2: PROBLEM IN INDEXING

The operation is that of boring 19 holes in the cast iron jig plate Figure 11.4(a), the holes being spaced as shown. Two set-ups are feasible: (1) with the dividing head set (Figure 11.4(b)) paralled to the machine spindle; (2) with the dividing head spindle vertical and the tool held in the vertical milling attachment. In case (1) the feed to the table is in a transverse

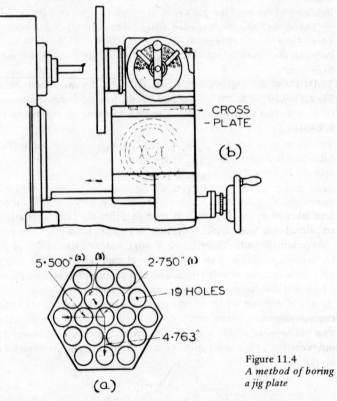

Figure 11.4
A method of boring a jig plate

direction, and in case (2) by vertical adjustment of the table. The first operation is to rough drill the holes and then to bore to size. Concentricity between centre hole 1 and the dividing head spindle is checked by a dial indicator held on the spindle of the machine or attachment.

Procedure

Bore centre hole 1 and then move the table 2.75 in. to the right to position the plate to bore hole 2 on the 2.75 in. radius circle. Position hole

3 on the same circle by indexing 40/6 = 6 turns and 16 spaces on the 24-hole circle of the index plate. When the six holes on this circle have been machined, the machine table is moved to the right 4.763 − 2.750 = 2.013 in. This locates the plate for machining the six holes on the 4.763 in. radius circle.

The six holes in this circle are spaced equally, but are displaced 30° with respect to the holes in the previous circle. Hence, to position hole 8 (the first hole on the 4.763 radius circle) in line with the tool, it is now necessary to index the plate 30° or the equivalent amount corresponding to 12 divisions, ie 40/12 = 3 turns and 8 additional spaces on the 24-hole circle. Hole 9 and each of the remaining holes are indexed by turning the crank 6 turns and 16 additional holes on the 24-hole circle, the same indexing as used on the 2.75 radius circle.

The same procedure is used when setting the place for machining holes 14 to 19 inclusive on the 5.5 in. radius.

Note that the indexing for hole 8, displaced at 30° from the previous hole, could have been obtained from the following: the angular indexing is 360/40 = 9° or 1 degree = 1/9 turn of indexing crank; therefore indexing for 30° gives 30/9 or $3\frac{1}{3}$, $3\frac{8}{24}$ as given when using a 24-hole circle.

EXAMPLE 3: HELICAL MILLING

To cut a helix on a milling machine, the dividing head is geared to the table screw so that, as the table is traversed, a rotary motion is given to the spindle of the dividing head. If equal wheels are mounted on the dividing head and the table screw a pitch of $\frac{1}{5}$ in. pitch, because 40 turns of the worm shaft of the head are required for one revolution of the spindle, the lead obtained is 40/5 = 8 in. (Note that this is 10 in. where the screw is ¼ in pitch.)

To find the gear train for any helix, the formula is

$$\frac{\text{lead of helix to be cut}}{\text{lead of machine 8 in.}} = \frac{\text{product of driven gears}}{\text{product of driving gears}}$$

Procedure

The procedure to cut a right-hand spiral gear with 24 teeth, a diametral pitch of 10 and an angle of 28.5° is as follows. The lead of the spiral, based on the pitch line diameter, can be found from

tangent of angle = circumference of work ÷
lead of spiral 0.543 = 3.14 × 2.2/lead = 12.722 in. lead

For a right-hand spiral, the right-hand end of the table is pushed towards the column of the machine, as shown in Figure 11.5, but this should not be done until the cutter has been set over the centre of the blank.

The gear train to give a lead of 12.722, from the formula given, is thus 12.722/8 or, from the wheels supplied, 32 × 48/ 40 × 24 or a similar ratio. These are mounted as shown in Figure 11.6.

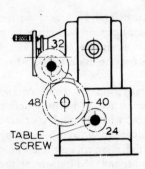

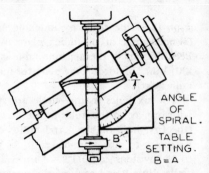

ANGLE
OF
SPIRAL.

TABLE
SETTING.
B = A

Figure 11.5
The mounting of gears for cutting teeth in a blank

Figure 11.6
A set-up for milling spiral gear

Spur gear cutters for milling helical gears are selected for the hypothetical number of teeth in the section at right angles to the helix rather than the actual number of teeth to be cut, and the formula used is

$$\frac{\text{number of teeth in wheel}}{\cos^3 C}$$

where C is the helix angle in degrees.

Having got the machine set up, the depth of tooth to be cut is obtained from 2.157/ diametral pitch = 0.216 in., while the index sectors for dividing for the 24 teeth require to be set to give 40/24 or $1^{16}/_{24}$ turns for each tooth.

After taking the first cut to about 0.180 in deep, return the cutter clear of the blank, and index for the next tooth, so continuing around the blank. Then sink the cutter to full depth, and repeat the operation. It is essential when spiral milling that a blank is tight on a mandrel, for the twisting action of the cut tends to slacken the work. Also, cutting should always be done towards the dividing head and not towards the tailstock.

EXAMPLE 4: FLUTING AN END MILL
For cutting the flutes of helical teeth end mills or twist drills and reamers, if special-shaped cutters are available the normal set-up is by mounting the blank between the dividing head and tailstock centres. If, however, special cutters are not available, a vertical spindle attachment can often be used with a standard cutter, as shown in Figure 11.7. Other helical work carried out in this manner includes the milling of helical splines or conical spirals on tapered milling cutters.

The example given is a six-fluted end mill, 1 in. in outside diameter, right-hand cutting with a 12° left-hand spiral and a flute angle of 95° with a corner radius of 0.031 in. The blank is mounted as shown and a 0.525 in. end mill with a 0.031 in. radius on the end teeth is held in the spindle of the vertical head.

The lead of spiral is tan (12°) = 3.14 X 1/ lead = 0.212 = 3.14 /lead or

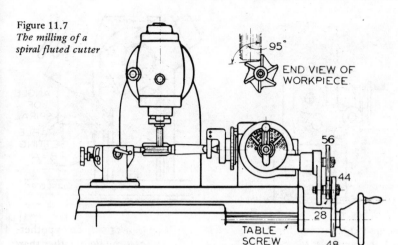

Figure 11.7
*The milling of a
spiral fluted cutter*

95°

END VIEW OF
WORKPIECE

56

44

28

48

TABLE
SCREW

the lead = 14.81 in., giving a required gear ratio of 14.81/8 (lead of machine) = 56 × 44 / 28 × 48 or a similar ratio mounted to connect the table screw and dividing head as shown.

As there are six flutes, the sectors on the dividing head index plate will be set to divide for each flute in turn, ie 40/6 = $6\frac{2}{3}$ turns, or 6 complete turns of the crank + 24 holes on the 36-hole circle. (Note that the 24 holes are in addition to the hole used for the index pin.)

The cutter is off-set from the blank centre by $\frac{5}{16}$ in. as shown in the end view of the work and, as left-hand flutes are required, the left-hand end of the table must be pushed away when the operator is facing the front of the machine and then must clamp at the 12° setting.

Procedure

As the cutter has a flat end, two cuts are required to produce the 95° angle, the first one giving 90° only when the cutter is set to depth and the first traverse made. The workpiece is then indexed for the next flute, and the operation is continued for all six flutes.

To remove the last 5° of metal to obtain the shape required, the blank is indexed by the crank handle 360/40 = 9 or as 1° = 1/9 turn; then 5/9 turn are required to bring the blank into position for the final cut; as the sectors are already set in the 36-hole circle, it is a simple matter to index 20 holes in this circle to locate the cutter to give the required angle of 95°. Thus, by machining and indexing as before for each of the flutes, the work is completed.

Assuming that the blank is of high-speed steel, annealed to 200 Brinell, a cutting speed of 150 ft/min is suitable, with a feed per tooth of 0.002 in. on the cutter in the vertical head. H.S.S. cutter.

EXAMPLE 5: GRADUATING (1)

The milling machine equipped with an indexing head can be used for graduating flat scales, verniers and other parts requiring odd fractional

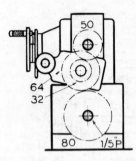

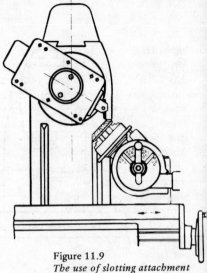

Figure 11.8
A gear arrangement for graduating

Figure 11.9
The use of slotting attachment for graduating

divisions or graduations. The spindle of the dividing head must be geared to the table feed screw, as in Figure 11.8, so that a longitudinal movement of the table is secured by turning the index crank. The driving gear is mounted on an arbor in the bore of the spindle, in a similar manner to when it is used for differential indexing.

By varying the indexing movement, graduations can be spaced with considerable accuracy. The graduation lines can be cut by a single-point tool held either in a fly cutter arbor mounted in the main spindle of the machine or between collars on the milling cutter arbor. The lines are produced by feeding the table transversely by hand, and the length of lines representing various divisions and subdivisions can be varied by noting the graduations on the cross feed screw.

If the gearing between the dividing head spindle and the table screw is equal, then 1 turn of the index crank will give a longitudinal table traverse movement of 1/40 × 1/5 (table screw) = 0.005 or 1/200 in. It is thus easy by utilizing the index plates to graduate any numbers divisible by five but, assuming that graduating lines 0.03125 or 1/32 in apart are required to graduate a scale, then the change wheels can be employed to vary the table traverse per one revolution of the index crank.

Thus for the dividing head spindle and table screw to rotate at the same speed the gear ratio should be 5 to 4, the pitch of the table screw being 1/5 in., and must be speeded up. The ratio 1/40 × 5/4 = 5/160 = 0.03125 in., suitable change wheels being 50/32 × 64/80 as shown in the diagram.

Thus the number of turns of the index crank for moving the table 1/32 in = 0.03125/0.00625 = 5 turns, the figure of 0.00625 being obtained from the total ratio of 1/40 × 5/4 × 1/5 = 0.00625.

If the divisions on a vernier reading to thousandths of an inch were to be 0.024 in. apart, the indexing movement would equal 0.024/0.00625 = 3.84 turns. This fractional movement of 0.84 turns can be obtained within close limits by indexing 26 holes in a 31 circle; thus 3 complete turns will give a longitudinal movement of $3^{26}/_{31}$ turns = 0.01875 + 0.00524 = 0.02399 in., which is only 0.00001 in. less than the required amount.

EXAMPLE 5: GRADUATING (2)

For graduating dials the vertical head or, as shown in Figure 11.9, a slotting attachment can be used in conjunction with the dividing head.

The dial blank can be held either in a chuck or on a stub arbor in the bore of the dividing head spindle. If the dial has a conical surface, the dividing head is tilted to the cone angle so as to present a flat surface for the single-point tool held in a short bar in the slotting attachment which is set for zero stroke. If the height between the attachment and machine table is limited, the attachment can be tilted to save the height required.

The knee of the machine is adjusted to bring the blank into light contact with the tool and is then raised 0.010 in. which is the depth of the graduations. With a 28° angle of the tool, the width of each graduation will be 0.005 in., which is sufficiently wide to be readily visible on the dial.

Assuming that 125 graduations are to be produced on the dial, every fifth graduation will be ¼ in. (16 mm) long while the rest will be $^{1}/_{8}$ in. (3 mm). Each graduation is indexed by moving the index crank 40/125 = 8/25. After cutting the first graduation ¼ in. (6 mm) long and retracting the tool, the crank is turned 8 spaces on the 25-hole circle. The work is then fed to the tool for a travel of $^{1}/_{8}$ in. (3 mm) for the intermediate three divisions, after which a ¼ in. (6 mm) travel is used for the fifth, and the procedure is repeated around the blank.

As this method requires some concentration in not overrunning some of the cuts, the alternative method is to index all around the blank with 125 divisions $^{1}/_{8}$ in. (3 mm) long and then to recut along each fifth division ¼ in. long, indexing 40/25 = $1^{15}/_{28}$ turns for each division. In many cases a still longer line is cut to indicate the tens divisions.

EXAMPLE 6: USE OF A SLOTTING ATTACHMENT

When equipped with a slotting attachment a milling machine can be used for cutting keyways or grooves, for working die sections or even for light shaping operations. Cutting speeds for roughing operations should be about 80 ft/min for cast iron, 100 ft/min for mild steel and 60 ft/min for alloy steels. Brass and aluminium can be increased to 150 ft/min. The amount of the feed per stroke should be from 0.002 to 0.003 in.

Machining the bore of a blanking die

The set-up for the operation and detail of the workpiece is shown in Figure 11.10. The bore is first drilled to provide an opening for the slotting tool, and the blank is then held in a chuck on the dividing head with the slotting attachment ram set in a horizontal position owing to height restriction if the attachment was used vertically. The tool is made of a triangular shape to suit the die contour, and the ram stroke is adjusted to ¾ in. to give some overtravel. The hand feed of the machine knee is used

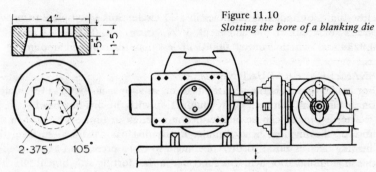

Figure 11.10
Slotting the bore of a blanking die

2·375″   105°

to apply the cut, the micrometer dial on the operating shaft being used to check for correct depth. An alternative method would be to set the edge of the tool central in the bore of the blank and to use the transverse feed instead of the vertical.

After cutting the first tooth space, the tool is lowered and the blank is indexed for the next one. There are 10 teeth to be machined; hence, the required number of turns of the index crank for locating each tooth will be 40/10 = 4 turns.

The cutting speed should be 60 ft/min with a hand feed of 0.002 in. stroke. After machining, work of this kind is often lapped by replacing the cutting tool with a lap held in the ram.

The attachment can be used in conjunction with vice work or with a circular table. By using the cross and longitudinal adjustment, operations such as the cutting of square holes in wrenches or other intricate work is easily carried out. Where specially shaped milling cutters are not available, teeth as in ratchet wheels can be cut, working in conjunction with a rotary table.

EXAMPLE 7: CAM MILLING

For this operation to produce a cam contour with a constant rise, the machine is set up as shown in Figure 11.11. The cam blank is held on an

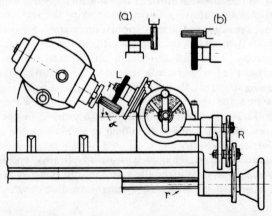

Figure 11.11
*The set-up for
milling a cam*

arbor in the dividing head, which is geared to the table screw by change wheels mounted to give the lead to the cam which is machined by the use of the vertical attachment. The dividing head is elevated and the spindle of the vertical attachment is set in alignment with the work spindle so that the cutter is parallel with the cam axis. The reason for elevating the dividing head spindle can be seen from the following. If the head is set vertical, as in Figure 11.11(a), a cam would be machined equal to the advance of the table for one cam revolution. If it is set as in Figure 11.11(b), ie the dividing head spindle is set horizontal, the lead would be zero and a concentric cam would be produced. Thus, by inclining the head to any intermediate position, the cam can be given any lead providing it is less than the lead for which the machine is geared.

Formula

To determine the angle of inclination for a given cam rise and with the machine geared for a certain lead, because $\sin A = R/L$ and $R = r/N$, we have

$$\sin A = r/N \times L$$

where A is the angle to which the dividing head and vertical attachment are set, r the rise of the cam in the given part of the circumference, R the lead of the cam, or rise, if the latter were continued at a given rate for one complete revolution, L the lead of spiral for which the machine is geared, and N the part of the circumference in which rise is required, expressed as a decimal in hundredths of cam circumference.

As an example, assume a cam is to be milled with a rise of 0.25 in. in 0.9 of the circumference, the machine being geared for 0.8; then $\sin A = 0.25/0.9 \times 0.8 = 0.347$ or $20° \ 20'$.

EXAMPLE 8: MILLING A DOUBLE-LOBE CAM

Object

The object is to produce a cam to the following specifications.
   (1) A constant rise of 0.225 in. for a rotation of $75°$.
   (2) A dwell through a rotation of $30°$.
   (3) A uniform fall of 0.225 in. for a rotation of $105°$.
   (4) A dwell through a rotation of $150°$

Equipment

The equipment consists of a Universal milling machine with vertical head and dividing head, and end mill 1 in. in diameter and a cam blank.

Theory

Let α be the angle of inclination of the dividing head, R the gear ratio between the lead screw and dividing head worm, n the number of threads per inch of table screw and L the lead of the cam to be milled.

For 1 turn of the table lead screw the dividing head worm will make R turns and, owing to a 40 to 1 reduction, the work will make $R/40$ turns. For 1 turn of the lead screw the table will advance a distance $1/n$ in.

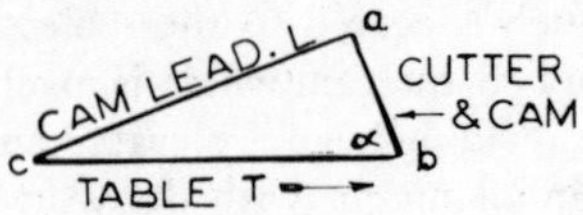

Figure 11.12
A diagram of the movements for cam milling

$$\frac{\text{turns of work}}{\text{movement of table}} = \frac{R/40}{1/n} = \frac{Rn}{40}$$

Therefore

$$\text{turns of work} = \frac{Rn}{40} \times \text{table movement}$$

or

$$\text{table movement} = \frac{40}{Rn} \qquad (11.1)$$

working on 1 turn of lead screw. From Figure 11.12

$$\frac{ac}{cb} = \sin \alpha$$

ie

$$\frac{\text{lead } L}{\text{table travel } T} = \sin \alpha \text{ or } T = \frac{L}{\sin \alpha} \qquad (11.2)$$

but, from Equation (11.1), $T = 40/Rn$; therefore

$$\frac{L}{\sin \alpha} = 40 \; Rn \qquad (11.3)$$

Calculations

For the specified cam, assume a gear ratio of 2 to 1; the lead of cam (specification, item (1)) is $L = 0.225$ in. in $75°$ or $L = 1.080$. $R = 2$, and $n = 5$ threads/in (fixed for machine). Thus

$$\sin \alpha = LRn/40 = 1.080 \times 2 \times 5/40 = 0.270 = 15° \; 40'$$

Similarly the uniform fall (specification, item (3)) can be found. Note that the dwells (specification, items (2) and (4)) will be milled with the gears disengaged.

Procedure

Mount cam blank on dividing head (Figure 11.11) and set up dividing head with any gear ratio 2 to 1 with lead screw. The cam blank can be marked out in $10°$ divisions, or with sufficient markings to assist in machining. In this case 75, 30 and $105°$. Check with a final $150°$. Note that 1 turn of crank gives 1/40 turn of blank or $9°$ turn. The cam will be cut in four settings.

(i) Incline dividing head to $\alpha°$ to horizontal and also incline vertical head $\alpha°$ to *horizontal*.

(ii) *Line* up scribed line with centre of cutter and put on a cut of 0.10 in.

(iii) Rotate dividing head crank and, with cutter rotating at 120 ft/min, commence cutting. Always let the cutter 'run out', the depth of cut decreasing.

(iv) Repeat operation (iii) until cutter is machining when blank rotates past the 75° mark.

(v) Disengage the gearing, line up the 75° mark with centre of cutter and turn dividing head crank to mill the dwell (specification, item (2)) of cam.

Finish milling cam but note that for specification, item (3), a reversing gear will have to be inserted in the gear train to produce a 'fall'.

EXAMPLE 9: MILLING CAMS BY INCREMENTAL SETTINGS

Many cams are made to a contour in which the incremental rise, for angular displacement, does not follow the law of the uniform rise cam. It is then no longer possible to obtain the profile by combining the uniform rotation of the blank with the table feed. Instead the cam is rotated through a small angle and is then displaced radially by a distance corresponding to the rise of the increment. This produces minute ridges, and the magnitude of the increments required to produce the contour determines the number of indexings.

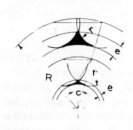

Figure 11.13
The ridges produced in cam milling

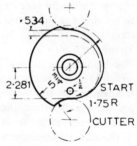

Figure 11.14
A radial cam with a 180° profile

The height e of each ridge (Figure 11.13) varies with the distance R from the cam centre, and angle c indexed in each setting, and the radius r of the milling cutter (also the radius of cam follower), as follows:

$$e = \frac{I\,R^2\,c^2}{8r}$$

where e is the height of the ridge over finished cam surface (11.14) in inches, R the distance from the cam centre in inches, c the angular displacement for each setting in radians and r the radius of the milling cutter, in inches.

For any selected height b of the ridges and radius r of the cutter, the number s of the settings divided by the radians is inversely proportional to the angular displacement c for each setting, ie

$$s = 1/c \qquad\qquad (11.5)$$

or combining Equations (11.4) and (11.5)

$$s = R\left(\frac{1}{8re}\right)^{1/2} \tag{11.6}$$

If e = 0.001 in., the maximum cam radius 4 in. and the minimum cam radius 2 in. with a cutter 0.5 in. radius, then the number s of settings per radian from Equation (11.6) is

$$s = 4\left(\frac{1}{8 \times 0.5 \times 0.001}\right)^{1/2} =$$

62.5 at the 4 in. cam radius and 31.25 at the 2 in. cam radius. Solving Equation (11.5) for c, and substituting the above values for s, it is found that, to keep the height of the ridges to within 0.001 in., the angular indexing for each setting should be 55' for the 4 in. radius, and 1° 50' for the 2 in. cam radius.

Radial cam with cam rise in geometrical progression

The cam profile extends through 180° (Figure 11.14). The rise, compared with the radial distance from the centre of the cam, does not require more than 3° angular displacement for each setting. Thus 60 settings are required. The blank can be held with the dividing head set vertically, or a rotary table with indexing can be used. A 3½ in. diameter end mill, equal to the diameter of the roller follower is required, and the indexing for each step of 3° = 40/120 or 10 spaces in a 30-hole circle.

Milling procedure

The blank is located with respect to the cutter by using the 0.5 in hole which locates the start of the contour. At this point the distance between the cam centre and the periphery of the cutter is 2.281 and, before mounting work and cutter, the two spindles of the vertical head and dividing head are aligned by moving the table saddle longitudinally, using test bars in the two spindles and a micrometer to check for the centre distance C. The dial on the table screw is now set to zero, while the centres of the blank and cutter corresponding to the low point on the cam is equal to 2.281 + 1.750 = 4.031 in. Hence, the additional movement the dividing head should move relative to the machine spindle in a longitudinal direction is obtained by subtracting C from 4.031 and, if C = 2.5, the table adjustment will be 1.531. When this adjustment has been made, the table dial is again set to zero.

A 0.5 in. plug is now inserted in the construction hole of the cam blank, and the low point of the cam is placed towards the cutter. It is then aligned with the axes of the cutter and cam blank. The first cut is taken by moving the table towards the cutter the same distance which it was moved away from the zero setting of the table dial. This will produce the first point of the cam profile located at 2.281 from the centre of the cam.

The blank is now withdrawn slightly from the cutter and is indexed for the second step 10 spaces on the 30-hole circle of the index plate; the table is then moved in a distance equal to the amount it was moved out,

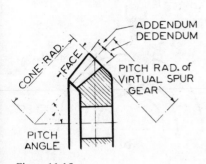

Figure 11.15
Nomenclature of bevel gear

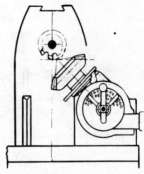

Figure 11.16
A set-up for milling bevel gear

less 0.004 in. to obtain the rise increment of the first step.

As a guide, an outline of the cam may be pasted or scribed on the blank, and a roughing operation can be carried out, keeping clear of the lines. As the rise is in a geometrical progression, it is necessary to tabulate certain positions for finishing, and the dial adjustment works out at 0.004 for position 1 after the start, 0.021 at position 10, 0.040 at position 20, 0.076 at position 30, 0.146 at position 40, 0.28 at position 50 and 0.534 at position 60. Intermediate positions may be calculated or judged from the scribed lines.

EXAMPLE 10: CUTTING A BEVEL GEAR

There are limitations on the accuracy obtainable when cutting the teeth of a bevel gear on the milling machine, and the process is not suitable for gears running at high speeds or for main drives, but satisfactory gears can be produced for feed motions or transmission connecting purposes.

The difficulty of cutting the teeth with a rotary form cutter is that the depth and thickness of the tooth reduces as it approaches the centre and, while depth does not impose much difficulty, special requirements are necessary to obtain the correct tooth thickness. As shown in Figure 11.15, the main dimensions are concerned with the thick end of the tooth, but it is the thin end of the tooth which limits the thickness of the cutter being used.

In order to choose the correct cutter it is necessary to determine the 'virtual spur gear', ie the size of a spur gear which would have the same tooth profile as on the edge of a bevel gear. This may be expressed as
pitch diameter of 'virtual spur gear' = pitch diameter X secant pitch angle
Hence the number of teeth in a "virtual" gear is $N_v = N$ X secant of pitch angle. Assuming a gear with 20 teeth, a diametral pitch of 4 and a pitch angle of $45°$, then 20 X secant $(45°)$ = 20 X 1.4142 = 28 teeth, or a number 4 cutter.

Procedure

The blank is mounted as shown in Figure 11.16, and the angle at which the dividing head should be set equals the pitch angle minus the addenum angle. Normally, the cutting angle for bevel gears equals the pitch angle

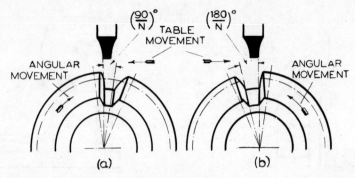

Figure 11.17
The operating procedure in milling bevel gear

minus the dedendum angle but, when using a formed cutter, the first is preferable as it gives a uniform clearance at the bottom of the tooth spaces and a closer approximation to the theoretical tooth shape.

The number of cuts required depends on the size of the bevel, for a large pitch; three cuts are usual starting with a central gashing cut, followed by a cut down each flank (Figure 11.17(a) and Figure 11.17(b)). Assuming the blank has been centrally gashed, it is now necessary to give an angular movement to bring the tooth flank radial, while the table must be adjusted to bring the blank into the correct position with the side of the cutter. The angular movement is equal to $(90/N)^\circ$ for the first flank. The off-set of the work is in the opposite direction to the angular movement, and the amount can be calculated from the following formula, assuming the cutter just cuts the correct width of gap at the small end:

$$\text{off-set} = \frac{C - F}{C \times \sin((90/N)^\circ)}$$

where C is the cone radius and F the face width (Figure 11.17(a)).

For cutting the other flank the angular movement is twice that of the first or second angular movement of blank equal to $(180/N)^\circ$, while the off-set is approximately double of that used for the first flank:

$$\text{off-set for second flank} = \frac{C - F}{C \times \sin((180/N)^\circ)}$$

(Figure 11.17(b)). For the angular movement, use the largest available circle, for conditions may arise where the index pin does not enter a hole in a small-number circle.

A PROJECT FOR DRILLING POLYGON SHAPES AND TOOLING EQUIPMENT

The production of special shapes on engineering components presents an

Figure 11.18
A tool for drilling ¾ in. square holes

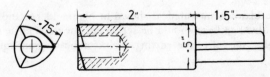

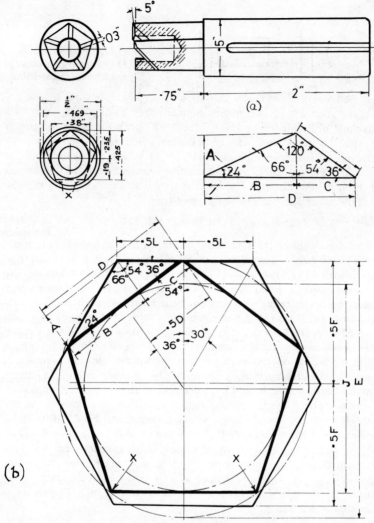

Figure 11.19
Drilling hexagonal holes

interesting study both for designers and for practical research work in machine tool laboratories in educational establishments. The subject is of increasing importance as new profiles for the Wankel engine, pump rotors and units for steam expanders continue to increase. Some of these profiles can be produced by a drilling action, the proviso being that either the drill or the work must float during the cutting action. The equipment described has been designed by the authors and has been used for demonstration in the Keighley Technical College.

Commencing with the two common shapes of polygonal holes, ie square and hexagonal, Figure 11.18 shows a tool for drilling ¾ in. square holes. With a tool of this type, if the centroid of the cutter is allowed to

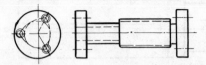

Figure 11.20
The machining operation to produce the square-hole drill

float, the corners of the tool with trace a square, the only slight inaccuracy being the rounding of the corners. Similarly, for drilling hexagon holes a pentagonal drill (Figure 11.19(a)) is used, the pentagon being the largest size capable of revolving inside the hexagonal guide bush.

Because of clearance at points X (Figure 11.19(b)), the figure produced has slightly convex sides, but the amount is almost negligible. In setting out the drill, the distance F across the flats is known, and from this the length of half of one side can be obtained from

$$L/2 = F/2 \tan (30°) = F/2 \times 0.577 \tag{11.7}$$

To find the length of the side of the pentagon to which the drill is made; from the left-hand triangle, $\sin (24°) / \frac{1}{2}L = \sin (120°) /D$. Therefore, from Equation (11.7), $D = \sin (120°) \times 0.288F/ \sin (24°) = 0.614F$. The dimensions next required are diameters I and E of the inscribed and circumscribed circles referring to the right-hand triangle, the first of these is found from $J = D \cos (36°)$. Substituting the value of D in terms of F gives $J - 0.614F \times 1.376 = 0.845F$. Similarly, the diameter E of the circumscribed circle in terms of F is given by $E = 0.614F \operatorname{cosec} (36°) = 1.044F$.

The machining operation to produce the square-hole drill is shown in Figure 11.20 and is by turning, the cutter being made about 1 in. longer than the finished length, and large enought at the ends to accommodate the three centre holes for turning the cam surfaces. The positions of the centres form an equilateral triangle with a perpendicular height of 0.649 in. The distance A will then be 0.649/3 or 0.216 in. and the distance B will be $0.649 - 0.216 = 0.433$ in. By mounting the blank between the three centres in turn, and running eccentric, radial sides are formed. After cutting off the surplus ends and drilling a $\frac{3}{8}$ in. centre hole, the teeth on the bottom can be milled or filed to resemble an end mill. The cutter is completed by hardening.

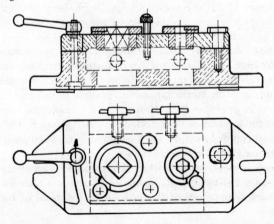

Figure 11.21
The jig used for experimental work in drilling square and hexagonal holes

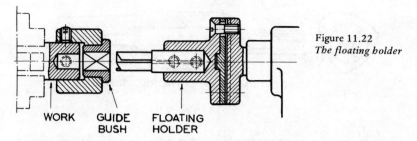

Figure 11.22
The floating holder

WORK   GUIDE   FLOATING
         BUSH   HOLDER

Figure 11.21 shows the jig used for experimental work in drilling square and hexagonal holes in steel plate ¼ in. thick. It is advisable to drill a round hole at first to reduce the wear of the polygonal drill, so two bushes with plain holes are used in the jig for the first operation, these being substituted by the bushes shown for the final operation. Using this jig necessitates a floating holder for the polygon cutting tools; an example is shown in Figure 11.22. This shows a simple method without the use of a jig. Assuming a square hole, say, a $^7/_{16}$ in. side is required in a shaft end. The bar is gripped in a chuck and a collar carrying a guide bush is made to fit on the bar end and to revolve with it. To save wear on the tool it is advisable to have a guide bush with a hole $^3/_8$ in. in diameter for a twist drill, and a second with a $^7/_{16}$ in. square hole to locate the cutter. The floating holder is made with a taper shank to fit the tailstock spindle, the floating action allowing the cutter to follow any position required by the square-hole bush revolving around it. The feed is by hand traverse to the tail-stock spindle.

A SQUARE HOLE WITH SHARP CORNERS

If this feature is required, the square hole in the guide bush must be larger than the required square while the cutter holder must be of a shape shown in Figure 11.23(a). Note that is is important that the relation between the cutter and the guide bush should be as shown. The tool is designed to fit into a drilling machine spindle, so a floating action must be given to the work. The example is for cutting $^5/_8$ in. square holes in chuck screws and Figure 11.23(b) represents the hole in the guide bush, 1½ in. square, with the cam which controls the cutting edge of the tool being shown inside. The cutting edge A is in the extreme corner of the square; this point is also the centre of the radius forming one corner of the cam. The distances AB, BC and CA are made equal to the side of the required square as before, but the point A is the only cutting edge which reaches the corners of the square.

The cam is shaped to make a continuous four-way contact in the guide bush, and one corner, as shown, must be an arc with the centre of the cutting edge. To satisfy the conditions, the cam must have four arcs, two drawn with A as centre, and two drawn from D and E respectively, each with a radius equal to the side of the guide square. Assuming that it is required that the square should have a side length of X, it can be seen that $R = AE$, and hence the side of the guide square can be obtained.

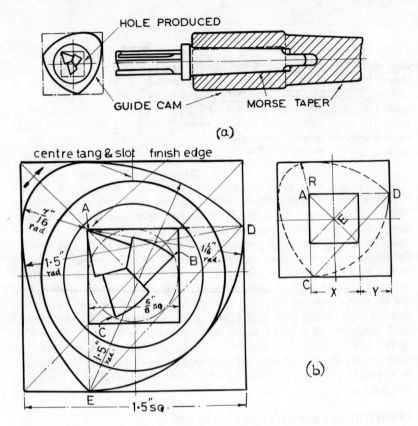

Figure 11.23
(a) *The shape of the cutter holder.* (b) *The hole in the guide bush*

Practical uses are for the holes in box keys, but a jig is shown in Figure 11.24, where purpose is the drilling of square holes in the ends of screws for lathe chucks. The screw is located vertically in a groove and held by a hinged clamp. A top bracket carries the guide bush and, to allow a floating action, the jig is made with a circular base. When on the drill table it rests on three steel balls to give free movement.

DRILLING HOLES OF PENTAGONAL AND TRIANGULAR SHAPE

A square-sided drill will produce a pentagonal hole (Figure 11.25). With one corner of the tool at X, two other points make contact with the side of the pentagon with two sides of the drill lying at an angle of $9°$ from the sides of the pentagon so that the angle $OXY = 54°$ and the angle $OXZ = 45°$ while angle $YXZ = 9°$ from the triangle XYZ; the length $XD = XY \cos (9°) = 0.987L$. The length $YD = L \sin (9°) = 0.156L$. The length $ZD = YD \tan (27°) = 0.156L \times 0.509$. The length S of the drill side equals $XD + ZD = 0.987L + 0.156L$; thus $S = 1.143L$.

In producing triangular holes the floating action must be greater than for other shapes described, the process of design being indicated in Figure

11.26. The length *L* of the side is taken to the sharp corners, although sharp corners cannot be produced on the workpiece but have a radius of about 10% of the length of the triangle machined.

From the centre line cutting the base at A, the distance AB is set out to equal ½*L*. Then from centre B and BA as radius the arc CAD is drawn. Using the same radius ½*L*, the arc CFD is drawn from point E on the extended centre line. Thus the section of the drill is that enclosed by the contour DFCA.

MACHINING OF CONJUGATE SHAPES

Many machine tool laboratories possess a gear shaper, and the object of this section of machining operations is to indicate how the use of the machine can be extended and how its installation can be justified. Practically any desired shape can be generated by applying a reciprocating motion to the generating point and moving it in a combination of directions. By the conjugate generating method, both tool and work have pitch line movement and this can be a combination of two pitch circles rolling together (as in normal gear cutting) or a combination of pitch circle and pitch line movement. With the first, both cutter and work rotate concentrically and maintain their respective positions when in contact with each other. In the second, the cutter rotates concentrically and the work moves in a straight line while maintaining the established centre distance at the full depth of cut.

TYPES OF CUTTERS

The shape of the cutter to produce any special component is not rigidly fixed but depends upon the sizes of the rolling circles on which the system

Figure 11.24
A jig used for drilling square holes in the end of screws for lath chucks

Figure 11.25
A square-sided drill for producing a pentagonal hole

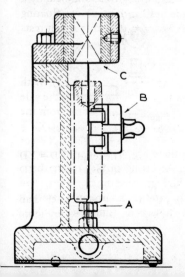

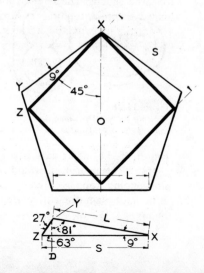

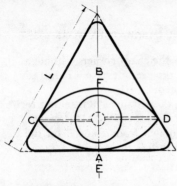

Figure 11.26
*The design of the floating action
for producing triangular holes*

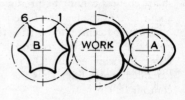

Figure 11.27
A profile

is based. Any changes in the sizes of the rolling circles affects the cutter shape so that profiles of identical shapes may be produced by cutters of different profiles but rolling on suitably proportioned rolling circles. For example the profile of Figure 11.27 may be produced by cutter A or by the intermediate shapes between A and B by a suitable choice of rolling circles and number of projections. Cutter B, although theoretically correct and capable of producing the correct shape, is not recommended, for much of the cutting would be done by the sharp points 1 to 6, which have a short life before regrinding.

Both cutter A and cutter B are of similar size, the rolling circles being indicated by the chain lines; however, A which more nearly approximates a circle has two repetitions of form whereas B has 6; the effect of this is taken care of by the change gears connecting the work and cutter spindles of the machine. Cutters with form relief instead of generated relief are easily made and, although such cutters are theoretically not correct throughout their life, errors are generally acceptable on most components.

Such cutters can be produced on the gear shaper; for example, Figure 11.28 shows a pump body to be machined around the outside edge. The

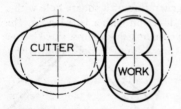

Figure 11.28
*A pump body to be machined
around the outside edge*

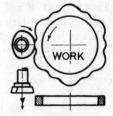

Figure 11.29
The production of an irregular cam

length of the major axis is 6 in. and, as there are no sudden changes in shape, a cutter of 4 in. would be suitable. A cutting surface rather than a single-point tool is preferable for machining constant-rise cams and, although a plain disc cutter could be used to produce an irregular cam as in Figure 11.29, radial control is required; however, using the moulding generating system and a conjugate cutter, the cam can be generated without radial movement of either work or cutter.

OFF-SET CONJUGATE GENERATING

The term 'off-set' is used when the cutter and work are of the same general shape with one member external and the other internal and when both members rotate at a ratio of 1 to 1, although the ratio is often varied as will be seen. Figure 11.30 shows examples of generating internal surfaces.

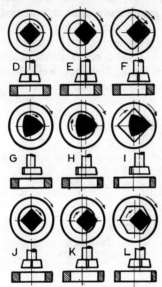

Figure 11.30
Examples of generating internal surfaces

Three steps are shown for each: (1) with the cutter and work on a common centre; (2) with the cutter fed into the work to the depth of cut; (3) the completed shape.

Figure 11.30, D, E and F, show the generating of a square hole with a square cutter, both rotating in the same direction with a 1 to 1 ratio. The same hole can be produced with sharper corners using a three-sided cutter as in Figure 11.30, G H and I. Cutter and work rotate as before but with a geared ratio of 3 to 4.

The generating of a hexagonal hole is shown in Figure 11.30, J to L, using a four-sided cutter. The cutter and work rotate in the same direction, but the ratio is 2 to 3. Figure 11.31 shows that a three-sided cutter will

Figure 11.31
The genration of a profile using a three-sided cutter

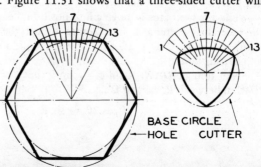

BASE CIRCLE
HOLE   CUTTER

generate the same profile, provided that the base circle of the cutter equals half the base circle of the hole and that the cutter rotates twice to one revolution of the work.

The examples given require no special work-holding equipment, but more complicated profiles can be generated with special fixtures. For example, there is a process known as 'describing' generating; this form reverts to the original concept of generating a surface by means of a point and differs from conjugate generating in that only one member has pitch line movement. The principle is especially applicable to cutting of clutch teeth where helicoidal surfaces are required. Figure 11.32 shows a clutch tooth component and, utilizing a single rounded point of one tooth on the cutting tool as the work is rotated, the cutter is advanced into the work to generate the correct form at full depth. The workpiece is located at a slight angle to allow clearance for easy withdrawal of the cutter.

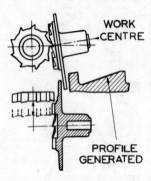

WORK
CENTRE

Figure 11.32
*A clutch tooth
component*

PROFILE
GENERATED

POLYGON TURNING BY FLY CUTTERS

Attachments are available for automatics, where as on most of the machines both the work and cutter spindle rotate in the same direction with a fixed speed ratio of two revolutions of the cutter spindle to one of the work. The action is shown in Figure 11.33(a), parallel flats being generated by single-tooth cutters, squares by 2-teeth cutters and hexagons by 3-tooth cutters.

The generation of a square is shown in Figure 11.33(b), the double cutter AP revolving about O at a speed $2N$ while the workpiece has a speed N in the same direction. While the face rotates from position AD to A′D′, the cutter rotates from PA to P′D′, owing to the speed ratio $2\psi = 2\phi$, however, $\psi = 90° - 2\theta$, hence $\phi = 90° - 2\theta$.

While the next face is moving into position, i.e. point D′ is passing to A, the second fly cutter at P′ must rotate through the angle P′OA = $180° - 2\phi$ = 4θ, and angle AWD′ = 2θ; hence the speed ratio is maintained during the non-cutting period.

From triangle OAW, $R \sin \phi = r \sin \theta$, but $\phi = 90° - 2\theta$; therefore the relation between R, r and θ is

$$R \cos 2\theta = r \sin \theta \qquad (11.8)$$

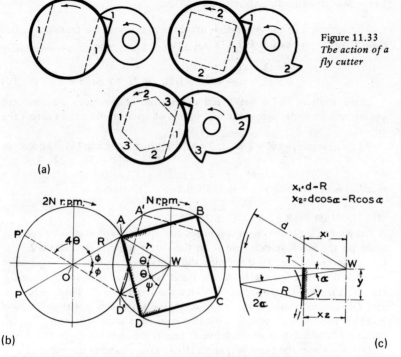

Figure 11.33
*The action of a
fly cutter*

(a)

(b)

(c)

Hence, if R and r are known, θ may be found from the quadratic equation

$$2R \sin^2 \theta + r \sin \theta - R = 0 \qquad (11.9)$$

The centre distance OW is given by

$$d = R \sin 2\theta + r \cos \theta \qquad (11.10)$$

To find by how much the generated surface differs from a true flat, in Figure 11.33(c) the work is assumed to be stationary and the centre line of length d to rotate at the same speed as the work but in the opposite direction. Also, at its extreme end the rotating line d carries a radius vector of length R. If, therefore, in generating half width of flat, the vector d has rotated through α, the vector R has rotated through 2α with relation to the vector d. The initial point T corresponds to $\alpha = 0$ and is at the mid-point of the flat.

Using the work centre as origin, the co-ordinates of the generating cutter are

$$x = d \cos \alpha - R \cos \alpha \qquad (11.11)$$
$$y = d \sin \alpha + R \sin \alpha \qquad (11.12)$$

If the point V corresponds to the extremity of the flat, or to a corner of the square,

$$\alpha = \frac{\psi}{2} = 45° - \theta \qquad (11.13)$$

The variation of the generated surface between V and T from a true flat is therefore, by equation (11.11) equal to

$$x_1 - x_2 = (L - \cos \alpha)(d - R)$$
$$= (L - \cos(45° - \theta))(R \sin 2\theta + r \cos \theta - R) \quad (11.14)$$

Since each of these bracketed expressions is positive, the resultant square will be wider across the mid-points of opposite flats than across the edges.

As an example, let $R = 2$ in. and $r = 1$ in.; then Equation (11.9) becomes $4 \sin^2 \theta + \sin \theta - 2 = 0$; whence $\sin \theta = 0.593$ and $\theta = 36° - 22'$. Also $\alpha = 45° - \theta = 8° - 38'$. Therefore by Equation (11.14) the deviation from a true flat is $(1 - 0.9886)(2 \times 0.955 + 0.805 - 2) = 0.008$ in.

MECHANISM FOR FLAT GENERATING

A flat-generating mechanism comprises two spindles connected by gears, A being on the work spindle and B on the cutter spindle; the ratio is 2 to 1. C is an intermediate. To determine the motion of a point in the plane of the gear B with reference to A, the latter is assumed stationary while the arm G (Figure 11.34(a)) rotates about the fixed axis O. Thus, while the arm OE rotates through the angle α, taking up a new position OE', the point in the plane B which originally coincided with O is moved to O', because of the rotation of B relative to arm G through the angle 2α.

O' lies on a straight line perpendicular to OE, whatever the value of α.

In practice, however, the flat to be generated does not pass through the axis of the work; hence the cutter edge does not follow a straight path. The curve produced may be investigated from Figure 11.34(b). Consider a point F in the plane of the cutter (ie gear B) originally at a distance b (*which is less* than 1) from the fixed centre O. The curve obtained commences at F where $\alpha = 0$, and when OE has rotated through α the

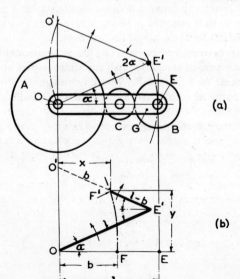

(a)

(b)

Figure 11.34
*The mechanism
for flat generating*

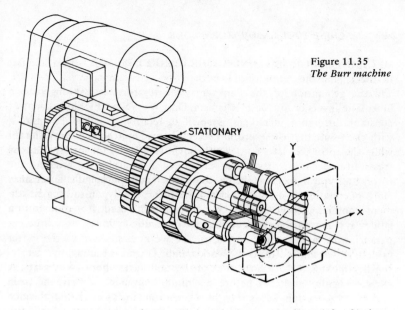

Figure 11.35
The Burr machine

STATIONARY

tracing point is at F'. From the figure, $x = b \cos \alpha$, and $y = (2l - b) \sin \alpha$. Squaring and combining these two equations and substituting for the terms $\sin^2 \alpha + \cos^2 \alpha = 1$, leads to the equation

$$\frac{x^2}{b^2} + \frac{y^2}{(2l-b)^2} = 1$$

This is the equation for an ellipse with semi-axis $2L - b$ and b. If the ratio

$$\frac{(2l - b)}{b}$$

is relatively great, the deviation of the generated curve from a straight line is correspondingly small.

MACHINING OF TROCHOIDAL CURVES

The term 'trochoidal curves' covers a range of closed mathematical curves which includes epitrochoids and their envelopes, the best known being employed in the engine of Felix Wankel, the Cooley engine and the well-known Geromotor pump mechanism widely used in the field of hydraulics. The problem in machining these special shapes arises from the fact that the rake angle of the cutting tool can vary from $+20$ to $-20°$ during rotation of the work, or tool, and it is this feature that necessitates a rather complicated mechanism to give a constant top rake cutting angle, usually by a rocking tool holder.

The Burr machine is equipped with a cutter head which has three single-point boring tools for machining the stator, as in Figure 11.35. The tools follow the path taken by the corners of the piston or rotor by the use of epicyclic gearing and eccentric spindle mounting which by crank arms causes the tools to swivel and to maintain constant cutting angles.

MECHANISM FOR MACHINING THE WANKEL PISTON

In the engine the rotor unit takes the form of the inner envelope curve, and the generation of the flanks may be regarded as resulting from a kinematic inversion of the mechanism. On a Wankel profile the amount of constant difference increment is small in relation to other dimensions, with the result that these profiles resemble true epitrochoids very closely while the rotor tips have a small radius which provides for sealing. Various types of mechanism have been developed for turning or grinding the rotors, a typical machine being the Pittler multiple-spindle automatic. Only one station is used for turning the rotor, a polygon turning attachment containing three equally spaced tools being used. It is mounted on guideways and movement is applied by hydraulic cylinders in a direction parallel with, and at right angles to, the work axis. Drive to the cutter spindle is transmitted from the work spindle (Figure 11.36(a)) by a single-tooth coupling R to ensure that piston and cutter head are set at the correct angular positions before machining commences. Drive between cutter and work spindles is then by a synchronizing mechanism S through gearing at T and then by a parallel-link coupling U.

Figure 11.36(b) shows the machining operation, the cutter head and workpiece rotation in opposite directions. The cutter is moved towards and away from the work by a cam V on the cutter shaft, the cam being held in contact with a follower by two hydraulic cylinders W.

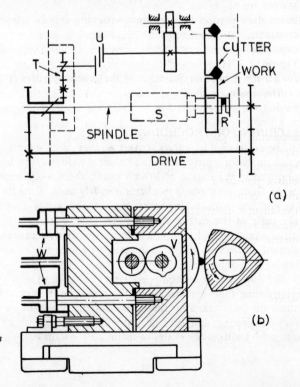

Figure 11.36
*The mechanism
for machining
the Wankel piston*

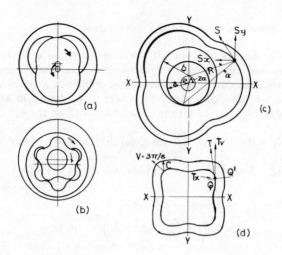

Figure 11.37
A profile

(a)

(b)

(c)

(d)

DEFINITIONS OF PROFILES

There are practical limits to the magnitude and direction of constant difference modification applied to epitrochoids and, if a large negative is applied to the profile in Figure 11.37(a) (Cooley engine), the resultant chords no longer resemble epitrochoids; the same applies to Figure 11.37(b) which is the Gerotor pump mechanism. An analytical definition of trochoidal curves has been developed at the Lakehead University, Ontario, Canada, by F.J. Robinson and J.R. Lyon, where rotors and stators are produced by NC on a milling machine, the units being for use in a steam expander. The curves may be defined in terms of four basic constants, the cartesian co-ordinates of points on the profiles being expressed as functions of these constants and a variable angular parameter. Figure 11.37(c) shows the geometrical relationship and angular parameter for a 3 to 4 ratio modified epitrochoid for the steam expander.

The basic generating circles have radii a and b, where a is the fixed circle for generating. If $b-a = e$ and $Ze = b$, where e is the eccentricity and Z a positive whole number, a set of unicursal curves may be generated according to the value of the radius vector R. Z is termed the 'winding number' of the curve and governs the number of cusps and lobes of the epitrochoid, while R prescribes which member of the family of epitrochoids is generated. Thus when $Ze = R$ the curve is an epicycloid. The increment c measured along the normal to the curve defines the profile, while the angular parameter α represents the rotor angle at any instant in a Wankel profile, and the co-ordinates of P' may be given as a pair of parametric trigonomical equations. Thus

$$x_p{}^1 = e \cos (Z\alpha) + R \cos \alpha + cS_y \qquad (11.15)$$
$$y_p{}^1 = e \sin (Z\alpha) + R \sin \alpha - cS_x \qquad (11.16)$$

where

$$S_x = \frac{dx/d\alpha}{((dx/d\alpha)^2 + (dy/d\alpha)^2)^{1/2}}, \quad S_y = \frac{dy/d\alpha}{((dx/d\alpha)^2 + (dy/d\alpha)^2)^{1/2}}$$

Point P$'$ describes a constant difference curve as α is varied through 2π radians and the origin of the cartesian system is at the geometric centre. Similarly, the envelope profile may be defined in terms of an angular parameter v relative to a cartesian system x, y, the origin at the geometric centre. The equations for the co-ordinate of Q$'$ on the envelope curve are

$$XQ' = R \cos (2v) - \frac{Ze^2}{R} \sin (2Zv) \sin (2v) \pm W \cos (2v) + cT_y \quad (11.17)$$

$$YQ' = R \sin (2v) + \frac{Ze^2}{R} \sin (2Zv) \cos (2v) \pm W \sin (2v) - cT_x \quad (11.18)$$

where

$$W = 2e \left(1 - \frac{Ze^2}{R} \sin^2 (Zv)\right)^{1/2} \cos (Zv)$$

and

$$T_x = \frac{dx/dv}{((dx/dv)^2 + (dx/dv)^2)^{1/2}}, \quad T_y = \frac{dy/dv}{((dx/dv)^2 + (dy/dv)^2)^{1/2}}$$

Equations (11.17) and (11.18) describe the complete envelope curve as the parameter v is varied through 2π radians, but for a given rotary machine only part of the envlope system will be required.

Figure 11.37(d) shows the true and modified envelope curves. The unit tangent vector T with components T_x and T_y defines the modified profile a distance c from the true inner envelope of the steam expander. The analysis holds for any values of Z, e, R and c, the profiles being computer plotted from Equations (11.15), (11.16), (11.17) and (11.18).

USE OF COMPOUND TABLES
Alternative machining methods can be used for both stator and rotor units. Tracer-controlled milling is a method in which a system provides simultaneous control of two motions at a constant feed rate, irrespective of direction.

A system for checking trochoidal forms has been developed by J.H.M. Ledocq, Belgium, the reference profile being produced under NC in conjunction with a computer. The workpiece is mounted on a rotary table, with the measuring probe provided with two additional movements to generate a non-circular form. Mechanical parts guide the movements, and the extent and relation depend upon data from either a tape or a small

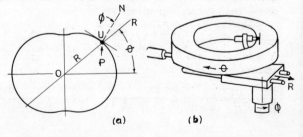

Figure 11.38
(a) *The geometry of a system for checking trochoidal forms.*
(b) *the lay-out for checking epitrochoidal curves*

(a)   (b)

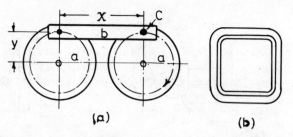

Figure 11.39
(a) *The principle of an arc-generating table.*
(b) *An example*

(a)

(b)

computer; the geometry is shown in Figure 11.38(a) where P is the probe which is displaced linearly along OR and is swivelled about the axis U. Rotation of the work about O is indicated by θ and swivelling of the probe by angle ϕ; successive settings of the probe in radial distances from O or OR, and the swivel settings about U, are related to the angular position of the work and result in the tip of the probe generating the theoretical trochoid.

Figure 11.38(b) shows the lay-out for checking epitrochoidal curves. The probe projects through the ring-shaped table while cylindrical guides carry the measuring head and allow it to move linearly in direction R. The swivelling movements ϕ are by the servo motor.

The subject is of increasing importance and a patented arc-generating table has been developed at the Admiralty Surface Weapons Establishment, Portesdown, to overcome the limitations of the usual methods of machining accurate profiles. The principle is shown in Figure 11.39(a) where two discs A are free to rotate about their centres and are connected by links B. Rotation of one disc is transferred to the other, and the radius of the pivots, as at C, can be adjusted so that the displacement of the link B can be varied. Further rotating members can be added and a wide range of contours produced. An example is shown in Figure 11.39(b); to produce the grooves on a milling machine requires four settings with possible mismatching, whereas, with the attachment, only one setting is required with the operation performed in one-fifth of the previous time.

PRODUCTION OF SPHERICAL PROFILES BY GRINDING

An interesting operation which supplements the turning of spheres in Chapter 4 is that of generating a sphere by grinding. The principle is shown in Figure 11.40(a) where, if a cup grinding wheel bored to a diameter D is set to a revolving workpiece at an angle α equal to half θ and is fed forwards until the point Y of the cutting line XY is coincident with the work axis AB, then a true spherical surface will be generated where $R = D/2 \sin \alpha$.

Since the cutting line is a circle, it will be apparant that a spherical surface will be generated whatever the setting angle, provided that the point Y is brought into coincidence with the work axis. The radius of the sphere will increase from $D/2$ to infinity as the setting angle decreases from $90°$ to zero. The proportion of a complete sphere of any specified radius that is generated will depend upon the diameter of the cutting line. A complete

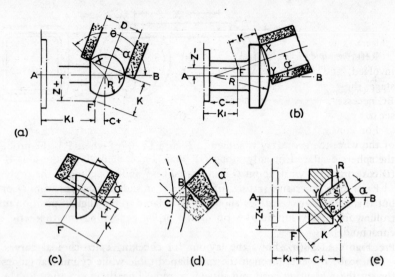

Figure 11.40
The production of spherical profiles by grinding

sphere of radius $D/2$ would result with a setting angle of $90°$, while a plane surface is produced at zero setting angle.

No oscillating motion has to be imparted to the grinding head; the axis F, about which the head is pivoted to the specified angle α, need not be located on the axis AB of the workhead, so that it is possible to locate the work in the grinding position if transverse and axial adjustments are available. The distance K1 from the face of the work spindle to the pivot point F is fixed, and the work is located axially by its position in the collet. The off-set Z of the workhead, and the distance $K_1 + C$ of the sphere centre from the datum face are calculated in relation to the datum distance K. Consequently, it is possible to mount a diamond for dressing the wheel at a datum position and to pre-set the axial in-feed stop for the wheel head.

The amount of off-set transversely from the pivot point F required for the workhead is given by $Z = K \sin \alpha - (D \cos \alpha)/2$. To find the position of the centre of the sphere from its relation to the spindle face, the distance $C = Z \tan \alpha$ is added to the constant K. It will be seen from Figure 11.40(b) that, if the setting angle is low, the off-set Z and the value C become negative.

With the arrangement shown in Figure 11.40(a) and Figure 11.40(b) the cutting is continuous with the edge of the wheel in full contact with the work, but it may be advisable to use a larger wheel at an increased centre angle α, as shown in Figure 11.40(c), so that contact is not continuous, for this method results in more efficient cutting and longer wheel life. Calculations must now be made with reference to the larger size of wheel and angular position chosen, and the in-feed movement applied to reduce the radius of the work by a specified amount is greater than that amount by reason of the angular setting of the grinding head axis.

Referring to Figure 11.40(d), because of the smallness of the amount involved, and the position of the wheel in relation to the work at this stage, the angle BAC can be assumed to be 90°. The amount of the in-feed BC necessary to reduce the work radius by an amount AC is therefore AC sec α.

For concave spherical surfaces (Figure 11.40(e)) the outside diameter of the wheel forms the cutting edge, and the distance from the centre of the sphere to the pivot point F is $K + PS$, and the off-set is $Z = K \sin \alpha + (D \cos \alpha)/2$.

Early in the grinding operation a land L, (Figure 11.40(c)) is formed, but this remains substantially constant; as the bore of the wheel is the controlling factor and this remains unaltered, the size of the sphere remains constant. Machines operating on the principle described are made by High Precision Equipment Ltd.

The datum dimension K is important in that it fixes the locating point for setting the work at the position of the truing diamond, while Figure 11.40(e) relates to the more general case of grinding and the calculations involved are $\sin \alpha = D/2R$.

A PROJECT FOR TRACER-CONTROLLED COPYING

CENTRE LATHE VERSUS COPY TURNING

This is a project for students in educational establishments which have a machine tool laboratory which includes a copy turning lathe. Practical examples of copy turning and producing the same part on a centre lathe are given with the tabulated times and graphs; these are of four components made on a lathe by T.S. Harrison & Son Ltd, Heckmondwike. It is suggested that for experimental purposes any suitable shaft or spindle is selected for the tests, using the same lathe throughout.

ASPECTS OF THE PROBLEM

First we must consider (1) the number of parts required and (2) whether the batch will be recurring at intervals. Assuming batches of 10 or less, the saving over machining on a standard lathe will be shown but, in a non-recurring case, the saving will be proportionally less owing to the cost of

Figure 11.41
The workpieces

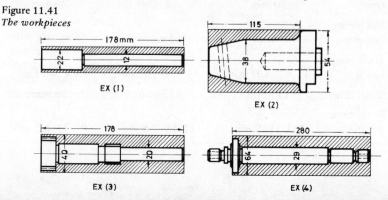

EX (1)

EX (2)

EX (3)

EX (4)

producing the master on a centre lathe before any saving can be obtained from copy turning. At some period, however, the master can take the place of one component in a batch. In both cases of recurring and non-recurring batches there is some allowance to be made for setting time.

The examples are all produced from Ledloy mild steel with the ends faced and centred and driven by a Kosta driver so that cutting could take place over the full length of the bar when required. The workpieces are

Table 11.1
Example (1)

	Standard centre lathe	Copying lathe	
		Non-recurring batch	Recurring batch
Setting-up time	20 min	30 min	15 min
Machining one piece	5 min	5 min	—
Machining batch	15 × 5=75 min	14 × 2.5=35 min	15 × 2.5=37.5 min
Total time	95 min	70 min	52.5 min
Total time per piece	95/15=6.3 min	70/15=4.7 min	52/15=3.4 min

Table 11.2
Example (2)

	Standard centre lathe	Copying lathe	
		Non-recurring batch	Recurring batch
Setting-up time	30 min	40 min	20 min
Machining one piece	7 min	7 min	—
Machining batch	15 × 7=105 min	14 × 5=70 min	15 × 5=75 min
Total time	135 min	117 min	95 min
Total time per piece	135/15=9 min	117/15=7.8 min	95/15=6.3 min

shown in Figure 11.41 and comprise the following: Example (1), a sliding wheel shaft; Example (2), a lever boss; Example (3), a rack pinion shaft; Example (4), a main spindle. Table 11.1 to Table 11.4 give details of the tests and the results obtained under the three following conditions: (a) normal centre lathe turning; (b) copy turning with a non-recurring batch; (c) copy turning with a recurring batch.

The examples are given in order of complexity, the tests being con-

Table 11.3
Example (3)

	Standard centre lathe	Copying lathe	
		Non-recurring batch	Recurring batch
Setting-up time	30 min	40 min	20 min
Machining one piece	9 min	9 min	—
Machining batch	15 × 9=135 min	14 × 15=70 min	15 × 5=75 min
Total time	165 min	119 min	95 min
Total time per piece	165/15=11 min	119/15=7.9 min	95/15=6.3 min

Table 11.4
Example (4)

	Standard centre lathe	Copying lathe	
		Non-recurring batch	Recurring batch
Setting-up time	40 min	50 min	30 min
Machining one piece	20 min	20 min	—
Machining batch	15×20=300 min	14 × 10=140 min	15 × 10=150 min
Total time	340 min	210 min	180 min
Total time per piece	340/15=22.6 min	210/15=14.0 min	180/15-12.0 min

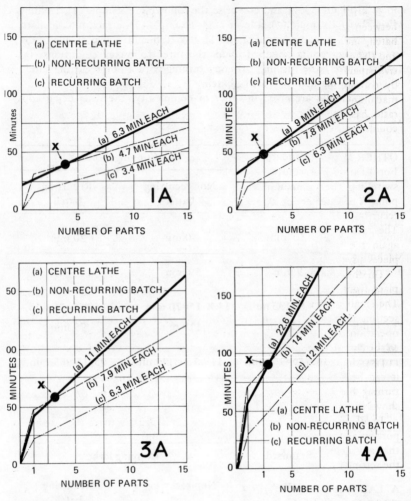

Figure 11.42
The time for setting up copy

ducted on a batch of 15 components in all cases; for the non-recurring batch it is necessary to produce the first one by ordinary turning methods and then, by using this piece as a master, to mount it for copy turning. This increases the time for setting up when copy turning and is indicated on the graph in Figure 11.42 at 1A, where the thin lines indicate the time taken for setting up and making the first piece. In making the first piece machining takes place simultaneously with setting up, but the increase in length of various lines indicates the additional time required to mount the master and adjust the tool. The remaining 14 pieces are then produced by copy turning.

It will be noted from the graph that the point of intersection X, between the curves of the centre lathe operation and the non-recurring batch, indicates that to produce four components the two methods (a) and (b) give equal times, but beyond this point the advantageous productive capacity of the copying system becomes apparent. This can be realized by considering the ratio of producing the first piece where setting up to machining time is 30/5 = 6 to 1 while, at the end of a batch of 15, the ratio has dropped to 30/70 = 0.43 (graph 1A, curve (b)). This feature could be plotted as curves of decreasing ratio values.

OTHER MACHINING FEATURES

For Example (2) the taper face was produced on the centre lathe by swivelling the compound rest. From graph 2A, curves (a) and (b), the point X of intersection shows that the advantage of copy turning on a non-recurring batch becomes prominent at five pieces made on the copy lathe. The rack pinion shaft (Example 3) includes no difficult contours, and again the intersection point at X of graph 3A indicates that after four pieces advantages accrue with copy turning on a non-recurring basis.

On the spindle (Example 4), for the first operation, machining takes place from the small end up to and including the face of the front flange. The small end requires bevels and recesses for thread rolling, and on this section the copy lathe scores heavily, and even more so in the second operation on the spindle nose which is more complex. Confining the test to the first operation, considerable metal removal is required, and from the intersecting point X between graph 4A, curve (a), and graph 4A, curve it can be seen that, even after three components, the advantage of copy turning begins to be indicated while the rapid divergence of the curves shows the similar rapid increase in production of the non-recurring batch.

In all cases curves (b) and (c) follow parallel paths after the setting-up portion as they obviously must do using the same lathe and conditions, so that any decrease in production time is simply obtained by a reduction in the setting-up time.

A LABORATORY PROJECT TO GENERATE AN INVOLUTE GEAR TOOTH

Object

The object was to demonstrate that a rack used as a cutter for producing gear teeth will generate a true involute curve.

Basis of experiment

The basis of the experiment was to establish the truth of the Sunderland system of gear generation, ie that as any gear of the same pitch will gear with a rack, a rack can therefore be manufactured as a cutter to produce gear teeth.

Apparatus

The apparatus consisted of a board and tee square with drawing equipment, including a template made in the workshop; this was a section of

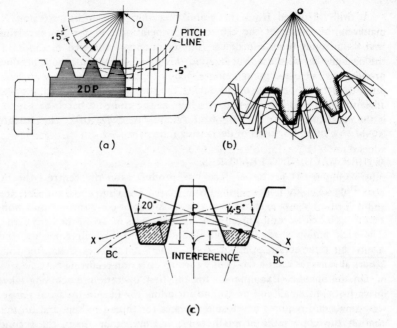

Figure 11.43
The generation of an involute gear tooth

rack of sheet aluminium with 4 or 5 teeth, a diametral pitch of 2 and a pressure angle of 20°.

Procedure

Figure 11.43(a) shows the procedure. Draw the pitch line diameter and the outside diameter of a gear with 14 teeth and a diametral pitch of 2 ie the pitch line diameter is 7 in. and the outside diameter 8 in. Using a quarter of the circle divide this into ½ in. spaces on the pitch circle diameter, and mark similar ½ in. spaces for the horizontal rack movement.

Place the template on the tee square just touching the gear blank and trace the portion of the template. Fix the gear centre O with a drawing pin so that the wheel diagram can be given rotary movement. Rotate the gear to the first ½ in. position and similarly move the rack by ½ in. distances, tracing the rack teeth after every movement along its length. As shown in the Figure 11.43(b) it will be seen that the cutter has produced spaces in the blank, leaving a tooth contour of involute form around the periphery; thus this proves the correctness of the Sunderland system of gear tooth generation. (Note that a fine pencil point must be used to contact the sides of the rack template; otherwise thin teeth will be drawn on the diagram.)

The project can be extended if a model rack section with a 14½° pressure angle is made for a second diagram (Figure 11.43(b)) shows the effect of interference and undercutting of the wheel.

A third diagram (Figure 11.43(c)) can also be drawn to show the maximum interference that will occur with a gear meshing with a rack and that the maximum addendum of the rack to avoid interference is equal to the distance between the interference point and pitch line of the rack. The proportions suggested are a few teeth of a 12-tooth wheel with a diametral pitch of 2 (a circular pitch of 1.571 in.); half the diagram has a pressure angle of 14½° and the other half has a pressure angle of 20°, as shown; X is the maximum addendum point on the rack to avoid interference which is equal to $D/2 - D/2 \cos^2 \psi = D/2 \sin^2 \psi$. Now, for standard gears, the addendum is the reciprocal of diametral pitch and thus 1/diametral pitch = $D/2 \sin^2 \psi$; however, from the definition of diametral pitch, $D = N/P$ and, substituting for D in the equation, $1/P = N/2P \sin^2 \psi$. Therefore $N = 2 \operatorname{cosec}^2 \psi$, where N is the minimum number of teeth to avoid interference and ψ is the pressure angle.

A PROJECT FOR PRODUCING PSEUDO-ELLIPTICAL GEARS

Elliptical gears find wide applications in textile machinery for the 'beat-up' motion on looms and on drawing machines to ensure the correct tension on yarns. Applications are also found on printing machines, on automatic wrapping machines, for tap changing on electric transformers, in certain computers and on machine tools. Indeed the increasing demand for these gears has led to the introduction of automatic gear cutting machines using either the hobbing or rack cutter process.

Object

The object is cutting of teeth in two elliptical blanks to show how the variable angular velocity of a driven shaft may be obtained from a driving shaft rotating at constant velocity.

Equipment

The equipment consists of a Universal milling machine with a dividing head and involute gear cutter of the formed type. Alternatively, a fixture with two slides at right angles can be used. To save cutting time during a demonstration, the blanks should be made of some soft material or plastic so that the cutting speed can be high.

Cutting the teeth

To simplify the cutting of the teeth of the wheels, one method is to generate an approximation to the elliptical curves by combining arcs of circles struck from selected points as centres, situated on the axis of the ellipse, so that the corresponding segments of the gears can be cut by rotating the blank about successive centres on a fixture adapted for use on a milling machine.

The acceptance of such a method of manufacture depends upon the divergence between the true elliptical curve and the closest imitation obtainable by a simple four-arc combination curve. Considering the distances M and N in Figure 11.44 and their geometrical relationship, the limiting values can be defined in the formation of an elliptically shaped

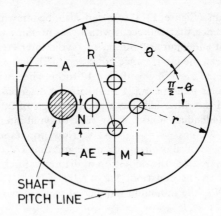

Figure 11.44
The production of pseudo-elliptical gears

figure described by combining four arcs of circles passing through the extremities of the major and minor axes of an ellipse. For any given ellipse of eccentricity e, the most appropriate relationship between M and N is $\dfrac{M}{N} = (1 - E^2)^{1/2} = \tan \phi$. This relationship leads to the determination of

the radii of the arcs in terms of θ; the larger radius is

$$R = A \frac{I - \sin \theta}{\cos \theta \,(\cos \theta + \sin \theta - I)}$$

and the smaller radius is

$$r = A \frac{I - \cos \theta}{\cos \theta \,(\cos \theta + \sin \theta - I)}$$

Having derived these formulae for the construction of an imitation ellipse, the resulting curve can be seen to lie slightly within the curve of the true ellipse from the extremity of the minor axis until the curves cross at some intermediate point, defined by an ordinate Y, where

$$Y = 2N \frac{1 - E^2}{E^2} - A \,(1 - E^2)^{1/2}$$

After this point the imitation ellipse lies slightly outside the true ellipse until it merges again at the extremity of the major axis.

The greatest distance between parallel tangents to the two curves is shown to be $\left\{(Y_e - Y_c + N)\cot \alpha + X_e - (X_c - M)\right\} \sin \alpha$, where X_e and Y_e denote the co-ordinates where the tangent touches the ellipse and X_c and T_c the co-ordinates where the tangent touches the circular arc; α denotes the angle made by the tangent lines and the major axis. At the point of maximum separation of the tangents in the end segment

$$X_e = \frac{A}{E} \left\{I - K^{2/3}\right\}^{1/2}. \quad X_c = \frac{A}{E} K^{2/3} \left\{I - K^{2/3}\right\}^{1/2}$$

where

$$K = \frac{\sin \theta - \sin \theta \cos \theta}{\cos^2 \theta \, (\cos \theta + \sin \theta - 1)}$$

θ is the angle whose tangent equals $\sqrt{(1 - E^2)}$ from the values of X_e and X_c (the points of maximum separation of the tangents), Y_e and Y_c and the values of α can be determined. From the formulae deduced for both segments it can be demonstrated that the divergence in the end segment lying nearest to the major axis is always the greater of the two.

Numerical calculations show that a pair of elliptical gears constructed for a 4 to 1 velocity ratio by the method evolved would have a pitch line divergence of 0.005 in. for every inch of maximum wheel radius whereas, if the velocity ratio were 1.5 to 1, the maximum error involved would be 0.0004 in. for every inch and would be negligible for gears up to about 10 in. diameter.

Since the error involved is dependent on the product AE and the velocity obtainable is dependent solely upon E, it follows that a low velocity ratio for large-diameter wheels may involve the same error as a high velocity ratio with small-diameter wheels.

If α denotes the angle between the major axis of the driving wheel and the line joining the centres of the shafts, V the velocity ratio and E the eccentricity, then we have

$$V = \frac{1 - E^2}{1 + E^2 - 2E \cos \alpha}$$

The maximum and minimum values of V may be derived by putting $\cos \alpha = +1$ or $\cos \alpha = -1$. Thus $V = \frac{(1 + E)}{\sqrt{(1 - E)}}$ for a maximum, and $V = \frac{(1 - E)}{\sqrt{(1 + E)}}$ for a minimum. Also, when $V = 1$, $\cos \alpha = E$. Now angular acceleration of

$$\text{the driven wheel} = \pm \frac{2E \, (1 - E^2) \sin \alpha}{(1 + E^2 - 2E \cos \alpha)^2} \, \psi^2$$

where ψ denotes the angular velocity of the driver. The maximum acceleration occurs when

$$\cos \alpha^2 = \left\{ 2 + \left(\frac{1 + E^2}{4E} \right)^2 \, \frac{1 + E^2}{4E} \right\}^{1/2}$$

P denotes the circular pitch of a standard form of tooth and N the number of teeth, then the contour of the pitch line suitable is given by

$$NP = 4\left\{ R\theta + r \left(\frac{\pi}{2} - \theta \right) \right\}$$

where θ is expressed in radians, $\tan \theta = (1 - E^2)^{1/2}$ and R and r denote the radii of arcs described from N and M respectively. Putting R and r in terms of α and θ,

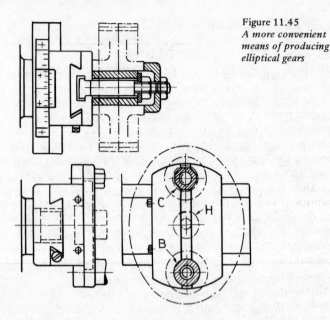

Figure 11.45
*A more convenient
means of producing
elliptical gears*

$$NP = \frac{4\alpha}{\cos \theta \, (\cos \theta + \sin \theta - 1)} \; \theta \, (1 - \sin \theta) + (\frac{\pi}{2} - \theta) \, (1 - \cos \theta)$$

Since tan $\theta = (1 - E^2)^{1/2}$ and the velocity ratio depends on E, the equation $NP = \alpha C$ may be written, where C denotes a constant appropriate to a particular maximum velocity ratio.

When a standard form of tooth is selected, it is impossible to predetermine the shaft centres and velocity ratio by giving them both precise values. Either the centre distance between shafts, or the velocity ratio involved in the constant C, must be subject to modification, but the departure from preconceived values will be small. A modification of centre distance amounting to about 0.01 in. will usually suffice to permit the use of a standard form of tooth, which is a desirable feature in manufacture.

Many elliptical gears are cut on milling machines without special equipment, for with a gear (Figure 11.44, for example) a hole at each of the four-arc centres enables the blank to be mounted by these holes in turn upon an arbor in the dividing spindle, but it is more convenient to have a fixture (Figure 11.45) consisting of two slides at right angles, one of which is adapted to fit the dividing head spindle, and the other one the work. The fixture is shown holding two gears through two bosses and the procedure in mounting the gears is with both verniers at zero, to centralize the button and to set the locating bush – at the calculated distance from the button by vernier and sideways by the test of a square on the machine table, the front slide being set vertically by a similar test. The indexing lever should be located with the plunger in the zero hole of the selected circle when the slides are at their right-angled position.

A bush C, cut away to avoid accurate radial adjustment, is lined up sideways with the button and the bush B, and the button is then removed. The bolts which hold the bushes also hold the blanks by separate clamping plates.

At every change of radius the centre of rotation must be moved to its proper position in line with the centre of the dividing head spindle; the main advantage of the fixture is that, once the blank is located as shown, the two slides allow adjustment for the four-arc centres without having to remove the blank from an arbor to another centre after part of the teeth have been cut.

Thus the procedure for gear cutting is to set up the blank with the centres of each arc in turn in line with the dividing head spindle, and then to divide for each tooth space by rotating the blank about the centre of the arc. The pitch angles, ie those subtended by the pitch ares, can be calculated, but it must be observed that, although the peripheral pitch is constant at every arc, the pitch angle varies with the radius of the arc. If the centre lines of spaces at the junctions of arcs of different radii do not fall exactly upon lines dividing the different sectors, the pitch length is made up of two arcs with different radii. Since angles under arcs of equal length vary inversely with the radii, at the change points the indexing angle will differ from that for spaces between the change points. Thus, as an example, to cut a gear of 42 teeth, the indexing may change in the following order of teeth, commencing at the top of the major axis: position (1), 4; position (2), 12, 1; position (3), 7; position (4), 1, 12; position (5), 4, = 42.

Figure 11.46
The use of an eccentric pinion as the driver

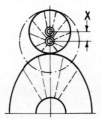

Note that the use of an eccentric pinion as the driver, as in Figure 11.46, simplifies matters. The wheel has twice as many teeth as the pinion, and for turning the pinion blanks it is only necessary to provide a mandrel with two sets of centres at a distance X, which is equal to half the throw. The off-set centres are used for turning the eccentric portion.

INDEX

METRIC AND IMPERIAL CONVERSION TABLES

The change from the imperial to the metric system was expected to be completed some years ago, but owing to reluctance to change in many sections of the engineering industry, both imperial and metric units are still being used.

As the teaching of the metric system predominates in education, students in engineering may find difficulty in converting metric to imperial units and the following table is given to assist in this purpose.

Linear measure
1 metre = 39.37 inches
1 centimetre = 0.3937 inch
1 millimetre = 0.03937 inch
0.001 millimetre = 0.00004 inch
(1 micrometre, μm)
1 inch = 2.54 centimetres
1 inch = 25.4 millimetres
1 micro-inch = 0.025 μm

Square measure
1 square metre = 10.764 square feet
1 square centimetre = 0.155 square inch
1 square millimetre = 0.00155 square inch
1 square inch = 645.2 square millimetres

Cubic measure
1 cubic centimetre = 0.061 cubic inch
1 litre (cubic decimetre) = 0.0353 cubic foot = 61.023 cubic inches
1 litre = 0.22 imperial gallon
1 cubic inch = 16.387 cubic centimetres
1 imperial gallon = 4.546 litres

Weight
1 metric tonne = 0.9842 ton (of 2240 pounds) = 2204.6 pounds
1 kilogramme = 2.2046 pounds = 35.274 ounces avoirdupois
1 gramme = 0.03527 ounces avoirdupois
1 ton (of 2240 pounds) = 1.016 metric tonnes = 1016 kilogrammes
1 pound = 0.4536 kilogramme = 453.6 grammes
1 ounce avoirdupois = 28.35 grammes
1 kilogramme per square millimetre = 1422.32 pounds per square inch
1 kilogramme/metre = 7.233 foot/pounds
1 pound per square inch = 0.0703 kilogramme per square centimetre

NOTES